A MATTER OF
LIFE AND DEATH

A MATTER OF LIFE AND DEATH

HUNTING IN CONTEMPORARY VERMONT

MARC BOGLIOLI

University of Massachusetts Press
Amherst & Boston

Copyright © 2009 by University of Massachusetts Press
All rights reserved
Printed in the United States of America
LC 2009028838
ISBN 978-1-55849-716-0 (paper); 715-3 (library cloth)
Designed by Jack Harrison
Set in Monotype Dante
Printed and bound by The Maple-Vail Book Manufacturing Group

Library of Congress Cataloging-in-Publication Data

Boglioli, Marc, 1969–
A matter of life and death : hunting in contemporary Vermont /
Marc Boglioli.
      p. cm.
Includes bibliographical references and index.
ISBN 978-1-55849-716-0 (pbk. : alk. paper) —
ISBN 978-1-55849-715-3 (library cloth : alk. paper)
    1. Hunting—Vermont—Addison County.
 2. Human-animal relationships—Vermont—Addison County.
         3. Addison County (Vt.)—Rural conditions.
    4. Addison County (Vt.)—Social life and customs. I. Title.
              GT5856.V4B64 2009
              799.29743'5—dc22
                 2009028838

British Library Cataloguing in Publication data are available.

Contents

Preface

When I turned nine I got a gun. As the official ninth birthday present for the boys in my family, it was one of the most anticipated events of my young life. In anthropological terms, it was a classic rite of passage. With this cherished gift I came one big step closer to the adult male world. And in my family this world still had a lot to do with classic indicators of American masculinity: physical conditioning, self-reliance, and, among other things, knowing your way around firearms.

Like those of my brothers before me, my first gun was a .177 caliber air rifle. Although an air rifle may not seem like a real gun, with only eight wobbly-armed pumps my Sheridan Blue Max could deliver a blow comparable to a .22 caliber rifle—or so I was told. Not surprisingly, gun safety was a serious subject in my house, and I dreaded the thought of my father learning that I had done something stupid with my gun. But along with the deadly seriousness that accompanied gun ownership came a level of trust and personal freedom that many contemporary American parents (particularly nonrural ones) would consider recklessly irresponsible. Yes, those were the days when a preteen New Jersey boy could roam the woods with his gun for an afternoon and his mother would be glad that he was out having some good clean fun.

For my first shooting lesson, my father, a former Marine Corps shooting champion, took me into our old dirt-floored wagon shed and set up a target against the fieldstone foundation. He showed me how to steady the gun by exhaling as I gently squeezed the trigger. After honing my marksmanship skills on targets, my next step was to start shooting live animals. That generally meant shooting "black birds," such as starlings and grackles. Killing a bird carried great importance for me. It was the next critical step in my adolescent march toward masculine self-reliance.

My opportunity came that spring. I stalked across the pasture from the barn to the old black walnut tree and huddled against its trunk, shielding myself from the grackle that was perched about thirty-five feet away, next to one of our chicken coops. I can't remember if I was standing or if I shot from one knee, but I do remember leveling the rifle with a smooth back-and-forth swing, exhaling, and squeezing the trigger—just as my father had taught me. The shot was perfect. When I inspected the grackle's body I found a bloody spot high under the wing where the pellet had entered. The bird had fallen the fifteen or so feet to the ground with nary a twitch. I swelled with pride.

As it turned out, the marksmanship was the easy part. Now I had to do something with the dead grackle that I held in my hand. I walked over to the wagon shed where, in addition to holding shooting lessons, we also kept our garbage cans. The only conclusion I had ever known to the process of killing nongame animals (groundhogs, rats, grackles, and so forth) was to throw them out—in the garbage, over a bank, or maybe even in the manure pile out behind the barn. What occurred next would prove to be one of the most significant moments of my life.

I tossed the grackle on top of the garbage and just stood there looking at it. My nine-year-old mind began to swirl. Only ten minutes earlier I was taking aim at my first kill with the steely resolve of a Louis L'Amour character, but now I was wondering why I had ever pulled the trigger. And it wasn't the actual death of the grackle that was bothering me. Growing up around farming, I was comfortable with the idea of killing animals. What was bothering me was one simple question: Why was this bird in the garbage? I'd practically have a funeral procession for a dead hamster, and this wonderful wild animal was headed to the town dump.

Sociologically speaking, I should not have been such a conflicted hunter. My father grew up on a hardscrabble farm as a hunter and trapper, my brothers hunted (one even trapped), my mom wasn't opposed to hunting, and I lived in a rural area. Nevertheless, after I killed that grackle my feelings about hunting were never quite the same. I wanted explanations. I wanted to really understand what people were doing when they went hunting. And, like many social scientists, I eventually sought out the answers to these deeply personal questions in my research. People often ask why I study hunting, and I usually just tell them that I've always been interested in understanding why people hunt and how different people interact with animals. And that is absolutely true. A more complete answer, however, would include the story of a nine-year-old boy staring at a dead grackle on a pile of garbage.

In the years between the grackle and graduate school I would do a little hunting, take wilderness survival courses, attend my father's funeral, raise pigs for slaughter, rescue and raise a newborn gray squirrel, spend dreamy moccasin-clad hours in the forests around our house, and generally fan my deep primitivist desire to live off the land. All the while my ambivalence toward hunting remained. But it had become more refined: my gripe with hunting was mostly with hunters, not hunting. Increasingly, I found myself in the ironic—but probably not uncommon—position of being pro-hunting but anti-hunter.

A prime example of my trouble with hunters occurred in the late 1980s or early 1990s as I was driving west on Route 22 in Whitehouse, New Jersey. A man was pulled over on the shoulder of the highway with a dead deer tied to the roof of his car. Another man immediately pulled over next to him and, as I was approaching in the oncoming lane, they exchanged a jubilant high-five. At the time, it struck me as one of the most disrespectful acts toward animals that I had ever seen, and it became symbolic of all that I thought was wrong with hunters: *they didn't care about animals!*

That moment also galvanized my interest in studying hunting, yet I never anticipated carrying out an academic study on the subject. For some reason, I assumed that high-minded professors would deem the topic far too mundane for serious intellectual scrutiny. It wasn't until my senior year in college, after my adviser at Rutgers, Uli Linke, assured me that hunting was a suitable topic for graduate studies, that I started to think about writing an academic book on hunting. Until then, I had dreamed of being a freelance writer and producing the Euro-American hunting version of Barry Lopez's *Of Wolves and Men*.

Finally, in the fall of 1996, after three years of graduate classes at the University of Wisconsin-Madison, I headed off to begin my ethnographic fieldwork on hunting in rural Addison County, Vermont. For the next eighteen months—in homes and small-town diners, over beers and on the trail—I made it my business to try to understand what hunting means to people in a rural community where hunting comes as "natural" as some people's daily visits to the local Starbuck's.

Like most anthropologists, I was treated to several crucial ethnographic moments during my fieldwork. These moments can be community-wide conflicts or seemingly mundane off-the-cuff comments made in passing. One of the most memorable occurred at a deer hunting camp in the Green Mountains during the second week of Vermont's 1997 rifle deer season.

Night had fallen and all the men had returned from the woods. They basked in the warmth of the camp, shedding their woolen coats and easing into their post-hunt routines. Some played cards, some sat on their beds fiddling with equipment, others started preparing dinner. I sat at a small wooden table with the card players—trying to learn a new game before my spare change ran out.

This was the third camp I had visited that year in search of data for my dissertation. As usual, I felt incredibly grateful to be at a camp. The Vermont firearm deer season is short, and hunting camps are enormously special places for many men. Considering that I had generally met the camp owners only once at their homes during the course of interviews (or, in one case, only over the phone), I was always quite impressed with the willingness of these men to allow an anthropologist to visit them while they were on vacation.

The camp I was staying at that cold November night was, and still is, located near Bread Loaf Mountain, not far from the site of the annual Bread Loaf Writers' Conference hosted by Middlebury College. Unlike most of the other camps I have visited over the years, this camp is not owned by one of the hunters; it is owned by Middlebury College and rented each year by a consistent core group. It is large and relatively modern by Vermont deer camp standards, but it was deer camp just the same: an easygoing, fun-loving, relaxing place where friends and family gathered every fall to joke about old times, ponder the mysteries of the natural world, and dream of future adventure.

As I will describe in detail in chapter 6, deer camps are eminently social places and, to a certain degree, the pursuit of deer simply serves as a good excuse to get folks together for a week or two in the woods. Hunting sets the tone and provides the reason for the gathering, but in the end it is probably the inner nourishment from deer camp's special camaraderie that keeps hunters coming back year after year, regardless of whether or not they kill, or even see, a deer.

So there I was drinking my beer and settling into life at a new camp when suddenly there was a wild-eyed man standing in the doorway. He seemed to bristle with excitement. Seconds later, I learned that he had come to let the guys know he had gotten a deer earlier that day. The man sitting to my right, a Middlebury native whose brother also attends this camp, gave me a quick nudge and encouraged me to go talk to the visitor.

After we'd been introduced and had talked for a little while, I followed this man out to his SUV to take a look at "his" deer's antlers. As I rounded the front of his vehicle, I was shocked by what I saw. There on the passenger seat, on an old newspaper, was the complete severed head of the deer. Dead deer are a common sight for me during my research; I've seen enough eighteen-month-old, one-hundred-twenty pound bucks to sink a ship. But for some reason, on that particular night, seeing a wide-eyed, bloody deer head sitting on a car seat seemed particularly shocking and, frankly, vulgar. I distinctly remember thinking about how important it must be for this guy to prove to his friends that he got a deer. To be completely honest, I was wondering if he was a nut. Or was he simply the personification of the "slob hunter" stereotype that informs so many non-hunters' opinions of hunters?

On the bright side, it was shaping up to be a beautiful night—the kind of night that the local Chamber of Commerce must dream of. The air was cold and crisp. The snow glistened under a clear, starlit sky. Smoke from the camp fireplace hung in the air, drifting ever so slowly across the distant face of Mt. Romance. Every now and then a tattered end of birch bark would rustle in the breeze.

The stranger and I talked for quite a while out in the driveway that night. As the minutes passed and I tried to will my teeth from chattering, the guy with the bloody deer head on his passenger seat seemed increasingly less scary. He told me how much he values his time in the woods, that he has a master's degree in Recreation, and how "complete" he feels when he kills a deer. I never did get to ask him what specifically he meant by "complete." But when the time seemed right I asked for his thoughts on an issue that I have discussed with numerous hunters over the years: the simultaneous violence and affection that is involved with hunting. He quickly said he knew just what I meant. Then he looked away into the shadowy woods—thinking. He turned back to me and shook his head, as if he were trying to put words to something he knew would be belittled by the constraints of our language. Finally, giving up on perfection, he shook his head again in what seemed like frustration and simply said, "Yeah, it's the ultimate paradox."

This relatively brief, chance encounter would prove to be thematically emblematic of my research in Vermont and, more generally, to the national discussion on hunting as well. People were killing animals they claimed to love. Women were taking to the field, guns slung over their shoulders, in unprecedented numbers. Over ten million Americans are hunting in the early years of the twenty-first century. It's not hard to see why contemporary American hunting strikes many as a paradoxical endeavor. Perhaps most importantly, however, this exchange beautifully encapsulates the deep, sometimes conflicted emotions that inform so many heartfelt conversations and bitter arguments about hunting in America today.

Acknowledgments

So many people have played a role in the completion of this book. First, I must extend my deepest gratitude to all the Vermonters who welcomed me into their lives and were willing to share so much. This book is a direct result of your generosity, and I dedicate it to you.

Several professors—both undergraduate and graduate—must be acknowledged for either their intellectual influences or more direct relationships to this book. Uli Linke gave me confidence that it should be written in the first place and has offered her theoretical insight while I wrote. Peter Nabokov shepherded me through the early days of graduate school, teaching me so much about so many different things—including anthropology. Tom Heberlein has kept my work in tension with the human dimensions of wildlife research in rural sociology and has offered more personal support through some tough patches than I could have ever expected or even imagined. Thanks again, Tom. Other professors whom I'd like to thank include Richard Flores, Neil Whitehead, Jack Kugelmass, Herb Lewis, Maria Lepowsky, Kirin Narayan, Paul Nadasdy, John Fiske, and Yi-Fu Tuan.

Many other colleagues and friends have made significant contributions to this process. Jan Dizard voiced his support of my work long before I had a book contract and offered detailed editorial advice on the manuscript that has significantly enhanced the quality of this book. Gerald Creed and Michael Harkin have provided invaluable intellectual commentary and professional advice. Adrie Kusserow has supported my career in various ways for a decade. Caitrin Lynch has been a trusted anthropological interlocutor. Richard Nelson sent a surprise postcard that bolstered my sagging spirits on many occasions over the years. I'll always feel indebted to Kristin Peterson-Ishaq for organizing my first talk at the Center for Research on Vermont. I also thank my colleagues in the Department of

Anthropology at Drew University for providing me with an intellectually stimulating home (and a junior academic leave!) to write this book.

Writing a book involves the contribution of many skilled laborers—only one of whom is the author. At University of Massachusetts Press, senior editor Clark Dougan provided support from the very beginning and patiently guided me to the places I had to be. My copyeditor, Mary Bellino, not only pulled me to the finish, but also sharpened my arguments with her insightful observations. Drew University undergraduate Jessica Glickman provided important assistance during the copyediting phase. I must also thank the talented staffs of the University of Wisconsin–Madison libraries, the Wisconsin Historical Society, the Vermont State Archives in Montpelier, the Vermont Folklife Center, and the Drew University Library. Lastly, I am grateful to all the Vermont Department of Fish and Wildlife personnel who spent hours of their precious time on the phone answering questions about wildlife management.

Thanks to Brian Troiano, Sara Farr, and Jonathan Blake for making my fieldwork so memorable and important; and to Pam, Ray, Eric, and Tyler for a great home. And thanks to all the folks from Woody's and Field Sports. I wish Rick Long were alive to accept my thanks for his warm and loyal friendship.

The support of my family was absolutely critical as I wrote this book. My brother Greg and his wife, Fran, were (and still are) indispensable to my research in Vermont. Through their hospitality and dependability I always have a place to sleep—even if I'm bringing my kids along now. My brother Peter has been a constant source of stimulating conversations about human–animal relations and animal behavior. My mother, Pati Boglioli, offered unconditional love and support (including financial) throughout my years in school and the research and writing of this book and I thank her from the bottom of my heart. I love you, Mom. And even though my father died twenty-five years ago, his passion for critical thinking and the outdoors have found their way into my life's work.

I cannot quantify my gratitude to my wife, Joslyn Cassady, for her emotional support, intellectual camaraderie, editorial eye, generosity, compassion, and belief in me. If it wasn't such a scary thought, I just might say I couldn't have done this without you, Jos. I love you. And I thank my precious daughters, Willa and Quinn, for being bastions of love, perspective, and hope. I love you both to outer-space!

A MATTER OF
LIFE AND DEATH

Introduction

If the human–animal boundary, and the parallel boundary that we like to draw between culture and nature, are as arbitrary a pair of constructs as evolutionary biology leads us to think, then the distinction between wild and domestic animals, between Wildtiere and Haustiere, is equally arbitrary. If so, it makes eminently good sense to see hunting as More's Utopians did, as just another species of butchery. And if we accept all this, then it seems hard to avoid coming to the same conclusion that Thomas More came to four hundred and eighty years ago: that butchery is not, in the final analysis, an appropriate recreation for a free people.

MATT CARTMILL, "Hunting and Humanity in Western Thought" (1993)

In 2001, while my wife and I were living on a Vermont island in Lake Champlain, we were invited to a dinner party by the local yoga instructor—a good friend of ours. Not long after taking our places at the dining-room table in her beautiful lakefront house, we were both asked to describe our doctoral dissertation research. The scene that played out was one the two of us were all too familiar with. Joslyn's research, based on nearly three years of fieldwork in the Arctic Circle with the Inupiaq (or Alaskan Eskimos), drew *oohs* and *aahs* from the guests, while my research in Vermont elicited only bemused silence.

One of the other guests was a recently retired English professor who had taught at a large northeastern university. When he heard what I was doing he raised his thick eyebrows and scoffed, "Bubbas." Although I had never before heard this term applied to hunters, I knew that it was a close cousin to that more common American antirural slur: *redneck*. I also took offense. My family has rural roots. And on my father's side these roots aren't only rural, but dirt poor as well. Hunting, trapping, and fishing played a legitimate role in his family's depression-era household economy.

After I briefly, and amicably, explained to the professor that most hunters aren't as bad as some would have you believe, he responded with a statement I found quite remarkable: "You're an academic. I'm retired. I can say these things." The obvious implication was that some sort of unspoken doctrine of political correctness prohibited me from speaking the truth about the people I had been studying intermittently for nearly five years.

Opinions like the professor's are not uncommon among the highly educated, urban elite who serve as our most influential producers of knowledge. In fact, I've become somewhat desensitized to them over the years. For example, it was Joslyn who pointed out just how strange, and significant, it is that almost every time I tell somebody about my research the next words I hear are "Do you hunt?" I can't remember how many times I've been given this obligatory moral litmus test. But nobody *ever* asks Joslyn if she kills wolves, whales, seals, or caribou like her Iñupiaq friends up north.

Although the urban elites would have you believe that their condemnation of mundane rural American practices such as hunting is a result of their highly developed ethical endowment, it's actually much simpler than that—virtually out of their control, in fact. It's primarily because they're urban people who engage in urban practices and have urban views of the nonhuman world. As Barbara Ching and Gerald Creed explain in the introduction of their fine and undervalued book *Knowing Your Place: Rural Identity and Cultural Hierarchy*: "In the West, few intellectuals have deep rural roots, and for those who do, education often severs these connections. The traditional pedagogical agenda, with its emphasis on enlightenment through the liberal arts, has long been opposed to the supposed essence of rusticity—lack of cultural sophistication and a preference for know-how over erudition" (1997, 10). In his foreword to a collection of essays titled *What Is an Animal?* the anthropologist P. J. Ucko makes a related point that has particular relevance for those of us interested in contemporary environmentalism: "Few of us who live in cities in the West, or for that matter our children, have any direct contact with animals other than certain variants of cats and dogs. Our conceptual stereotypes of humans and animals derive largely from television, nature films, and books" (1988, xii).

Indeed, I would go so far as to suggest that many urbanites are not culturally prepared to deal with the topic of hunting, at least initially, in anything but an ambivalent, if not adversarial, manner. Census reports and sociological studies repeatedly show that urban and rural people often engage in different outdoor activities (bird watching versus hunting, for example) and express vastly different sensibilities regarding the nonhuman world (sometimes phrased as *aesthetic* versus *utilitarian*). Killing beautiful wild animals simply does not fit into the mainstream urban worldview. This is why Vermont's Department of Tourism and Marketing does not aggressively promote hunting—an industry that brought nearly $200 million in expenditures to the state in 2006 (USDI 2006b, 15). There is a real fear that talk of hunting in promotional materials might scare off some of the people who help make Vermont tourism a $1.5 billion industry that accounts for approximately 12 percent of the state's jobs (VDTM 2007, 2, 12). In the meantime, Vermont will continue to serve up a warm and fuzzy theme-park rendition of rural life—touristic pablum for people visiting from New York, Boston, and Montreal. (A Department of Tourism and Marketing employee told me about a potential foliage viewer from Britain who called their office and to ask if she would have to deal with seeing deer carcasses strewn about the woods. She was reassured that she would be arriving before hunting season.)

Not surprisingly, as the United States becomes more urbanized, rural sensibilities become increasingly marginalized. On the heels of marginalization follows the delegitimazation of rural practices and the ascension of urban environmental sensibilities to the dominant status of "common sense" in mainstream, media-driven American culture. In this political climate, hunters often become the subjects of emphatic ethical critiques. Take, for instance, freelance journalist Joy Williams's remarks on hunting and hunters in *Esquire*: "Sport hunting is immoral; it should be made illegal. Hunters are persecutors of nature who should be prosecuted" (1990, 128). Even more arresting is Matt Cartmill's characterization of hunting in his acclaimed book *A View to a Death in the Morning* as "the rural equivalent of running through Central Park at night, raping and murdering random New Yorkers" (1993, 239; also quoted in Dizard 2003, 39). Or consider an extended commentary on rural hunters by the philosopher Tom Regan, author of the influential book *The Case for Animal Rights*: "There's a lot of anthropology going on here, and there's a lot of sociology, a lot of class stuff, a lot more than a philosophical argument. The whole gestalt of rural people, their whole way of viewing the world is radically different from somebody living in Washington, D.C., and the more we understand the other layers of it, the more we can practice 'hate sin love the sinner.' Even if we disagree with the other person, we must affirm that person's humanity, and not get to the point where we can confuse that person's character with that person's actions. Good people can do bad things. Even if we disagree morally with them about what they're doing, many of them are still good people. That's a tough thing for the animal rights people to keep in mind, but I think it's true" (quoted in Kerasote 1993, 263). Despite his empathetic tone, Regan ultimately characterizes hunting as a morally deficient aspect of rural American culture.

Another, more subtle example of the villainization of hunters comes from an article in *Backpacker Magazine* that warns readers of the dangers of getting too close to wild animals: "While some people believe animals are a resource put on Earth to be used by humans—hunters and farmers mostly—there are those at the other end of the spectrum who view beasts as cute, petlike. . . . So the next time you're extended the privilege and honor of sharing land that's home to wild creatures, hold them in high regard, respect their ways, and teach your fellow travelers to do the same. Remember there's a reason we don't call them *tamelife* and that they have a right to exist free of human interference" (Shealey 2000, 14). Apparently, this author has no qualms about stereotyping 12.5 million American hunters as folks who neither hold animals in "high regard" or "respect their ways." Certainly I would never claim that being a hunter makes anyone an angel, but this kind of blanket statement is patently absurd. In fact, one of the most important lessons of my research in Vermont has been that the exploitation of animals does not stand in the way of a hunter's feeling respect, or even affection, for those animals. In this regard, the perspectives of Vermont hunters represent a dramatic departure from what I have heard in both interviews and informal conversations with urban nonhunters. Many meat-eating nonhunters openly admit that they cannot psychologically cope with the idea that their food is actually the flesh and blood of another animal, and they say that they appreciate the fact that meat is packaged in cellophane and Styrofoam so it looks nothing like the animal's body from which it was taken. Likewise, vegetarians often turn ashen when they learn that the corn or soybeans they eat could not be grown without killing prodigious numbers of white-tailed deer. As Richard Nelson, a cultural anthropologist who has written widely on the topic of hunting, writes in *Heart and Blood* (which examines the relationship between humans and white-tailed deer), "Whenever any of us sit down for breakfast, lunch, dinner, or a snack, it's likely that deer were killed to protect some of the food we eat and the beverages we drink" (1997, 310). Nelson argues that this ignorance of, and discomfort with, basic facts of food production is often a product of urban living: "People like ranchers and farmers, whose earthbound labors feed the neighboring city folk, often share a similar love for the land's beauty, but they also work on it, use what it provides, and live in direct physical reciprocity with it. Because of this, I believe, they often have a fuller sense of connection with the surrounding environment, not just as scenery but as their own living source, a world that includes and wholly engages them" (1997, 177).

What's more important, however, is that the *Backpacker* author I quoted earlier explicitly contends that *human beings should not be disturbing wild places.* This common tendency of environmentalists to define wilderness as those places that haven't been affected by humans is at the root of much antihunting sentiment because, ultimately, it leaves no ethical way for nonindigenous people (even the most strident animal protectionists usually reserve the right for indigenous peoples to hunt) to kill wild animals. Not to mention the fact that the fantasy of the pre-Columbian wilderness depends on discounting Native Americans as humans who were inhabiting the North American landscape.

Today, however, we know that "wilderness" is more of a philosophical space than an actual geographical place, and a number of people have written eloquently on this issue. Most noteworthy, perhaps, has been the work of the environmental historian William Cronon.[1] In his influential 1995 article "The Trouble with Wilderness," he writes: "The dream of an unworked natural landscape is very much the fantasy of people who have never themselves had to work the land to make a living—urban folk for whom food comes from a supermarket or a restaurant instead of a field, and for whom the wooden houses in which they live and work apparently have no meaningful connection to the forests in which trees grow and die" (80).

Regrettably, many intellectuals (including members of my own field of cultural anthropology) have been slow to appreciate both the complexity and diversity of human–nature relations among Euro-Americans. Often glossed over in one-size-fits-all commentaries on so-called Euro-American or Western human–nature relations (what I will call "environmental occidentalism"), these descriptions often distort the realities of *specific cases* and certainly do not provide many possibilities for finding similarities between the ways that, say, a subsistence hunter/gatherer in the Arctic and a Vermont dairy farmer learn about, understand, explain, and interact with the flora and fauna around them.

Even more difficult for the intellectual community has been coming to grips with *rural American* human–nature relations.[2] As the historian Karl Jacoby argues in *Crimes against Nature: Squatters, Poachers, and the Hidden History of American Conservation*, rural Americans play the singular role of ecological villain in dominant historical narratives about the birth of the American conservation movement, a position that essentially argues, or certainly implies, that market hunting and ecologically reckless subsistence hunting was the norm among nineteenth-century rural Americans.[3] Wealthy, highly educated, politically connected *urban* sporthunters like Teddy Roosevelt, on the other hand, are given the plum role of noble ecological savior. As Jacoby writes, "Ever since Marsh's *Man and Nature*, a key component of conservation's degradation discourse has been the need to use science and the state to protect nature from the recklessness of rural folk" (2001, 198).

The tendency to overgeneralize about the ways rural Americans engage their physical surroundings has been a frustration of mine for some time, and it is the source of my gratitude for wonderful historical studies such as Jacoby's *Crimes against Nature* and Louis Warren's *The Hunter's Game: Poachers and Conservationists in Twentieth-Century America*. While commentators generally acknowledge that rural areas are the crucible of American hunting (because, even if most hunters do not actually live in rural areas, the proportion of people who hunt in rural areas is far greater than in urban areas; because hunting usually takes place in rural areas; and because hunting is an unquestionably more in line with rural culture than urban culture), actual contemporary rural hunters are too rarely asked to share their perspectives by people who write about hunting. And when contemporary hunters' perspectives are discussed, we are often treated to a quote from a very peculiar type of hunter: the kind who writes books about hunting. Regardless of how good they are at hunting and writing about hunting, folks like Richard Nelson, Ted Kerasote, James Swan, Jim Posewitz, and Pam Houston are hardly representative of American hunters. This book is an attempt to paddle upstream against this dominant current by applying the traditional ethnographic method of cultural anthropology to a topic that, heretofore, has often lacked real human faces and voices. And I want to emphasize that I am focusing on rural Vermont hunters—not jet-setting trophy hunters in search of obscure species in faraway places, but the kinds of people who really constitute the heart and soul of American hunting: middle- and working-class white people who do most of their hunting close to home.

Through the example of hunting in rural Vermont, I hope to demonstrate that, early in the twenty-first century, there are Americans who have successfully managed to both consume and respect the animals in their presence. To my mind, many of the rural Vermonters I know are succeeding at what Cronon calls "the unending task of struggling to live rightly in the world" (1995, 90). More pointedly, although I will not spend much time hashing out a detailed philosophical argument, I believe my research clearly shows that antihunting sentiment is not a product of a more highly developed sense of ethics or a greater "respect for nature" among non-hunters, but rather is just one aspect of a rural/urban cultural debate that masquerades as a discussion of moral absolutes.

There are approximately 12.5 million hunters in the United States, representing 5 percent of the country's population (USDI 2006a, 65). There are more hunters than downhill skiers, horseback riders, or backpackers. There are more hunters than cross-country skiers and snowmobilers combined (Cordell et al. 1995). In fact, hunting is more popular in the United States than in any other "developed" nation (Heberlein, Ericsson, and Wollscheid 2002). There is, however, one critical difference between hunting and virtually all other outdoor recreational activities: in the eyes of many Americans hunters are *moral suspects*.[4]

In organizations such as People for the Ethical Treatment of Animals (PETA) and the Fund for Animals, the depravity of hunters is a ubiquitous topic of discussion. Whether drawing on well-worn stereotypes of Joe Six-Pack (the archetypal "slob hunter"), or mounting philosophical assaults taken from the pages of Peter Singer's *Animal Liberation*, members of these groups commonly portray hunters as morally corrupt, insensitive to "nature," infected with a wanton desire to kill—indeed, quite possibly murderers (Singer 1990). As the sociologist Jan Dizard notes, "Allegations of bad character abound in the literature critical of hunting" (2001, 16).[5]

Suspicion about hunting, however, is not confined to the ranks of high-profile animal rights organizations. Numerous questions circulate in everyday discourse that concern not only the morality of hunters, but also the effect of hunting on society at large. Does hunting lead to aggression against humans? Is hunting an expression of hypermasculine domination? Do people hunt for the joy of killing? Does gun ownership cause violence?[6] Do the efforts of the National Rifle Association to keep gun ownership legal actually put illegal guns into the hands of those who would commit violent crimes? Is hunting a necessary, or effective, wildlife management strategy in the twenty-first century? These kinds of questions, as well as the growing influence of environmentalism and animal protectionism, have generated a considerable amount of ambivalence among the non-hunting 94 percent of America. I recall a woman who was originally from a large city, but was then living in rural Vermont raising organic mushrooms, exclaiming something along the lines of "How can you stand to be around those people?" Apparently she assumed that my graduate advisors had *forced* me to travel to Vermont and study hunters, since clearly nobody who could mix a good martini and was so highly educated would actually *choose* such a miserable fate. She was also disgusted that women hunt, as if it were the ultimate desecration of female nature. It is easy to understand why Dizard predicts that "hunting will edge nearer and nearer the center of our 'culture wars'" (2001, 23).

The ways these questions are dealt with in environmental and political circles will have important consequences for many sectors of American society. One could argue that the most far-reaching effects, however, will be felt among rural white Americans.[7] It is well documented that hunting plays a prominent, sometimes even celebrated, role in rural life and is important in the development of community solidarity and rural identity (see Fitchen 1991; Marks 1991; Nelson 1997; Nemich 1996; Muth and Jamison 2000). Regrettably, the meanings and importance of hunting to many rural individuals, families, and communities is often unappreciated by the majority of Americans who live urban lives and have little personal association with rural hunters.

The primary goal of both my initial and my continuing research has been to learn about the ways hunters in rural Vermont conceptualize their relationships with the animals they hunt. Questions such as these have guided my research on this front: Are there any shared ethical guidelines for the killing of animals? Do hunters like animals, or is hunting based in human–animal antagonism? What are the emotions associated with killing animals? As this book unfolds, it will become clear that the ways that hunters in Vermont relate to animals cannot be understood without taking into consideration the everyday experiences of rural life, the specific historical events that have shaped the state of Vermont since the colonial era, and hunters' general philosophical views on one exceedingly complex issue: "nature."

Another area of inquiry that I considered absolutely essential to my research was gender. After all, less than 10 percent of the hunters in the United States are female. During my fieldwork, I spent many hours talking to people about this subject as I attempted to learn how men felt about the appropriateness of female hunters, whether hunting should be considered a performance of hypermasculinity (or even misogyny), and why more females are hunting at a time when hunting is increasingly thought of as a masculine anachronism. More specifically, I was interested in comparing male and female *hunters*, not just learning about why more men hunt. My central guiding questions here were: Do men and women share common sentiments about their relationships with the animals they hunt? Are all-male deer camps in Vermont comparable to the "men's houses" in faraway places like Papua New Guinea? Are female hunters highly politicized feminists who are trying to "change the world," or are they (as many people have asked me) "trying to be guys"?

My third general area of concern was the relationship of hunting and modernity. Considering that hunting is an archaic human practice that is scarcely necessary in contemporary America, how does hunting manage to remain socially relevant and personally meaningful? Does hunting somehow ease the pressures of modern life in ways that other "recreational" activities do not?

The ethnographic fieldwork I undertook in order to address these questions began in the fall of 1996 in rural Addison County, Vermont. Virtually any place in the state of Vermont, apart from the more urban Chittenden County, would be an excellent location for a study of hunting.[8] Among U.S. states, Vermont has the lowest percentage of population living in urban areas (U.S. Census Bureau 1998, 46, and 2000a, 38), a high hunting participation rate by national standards, at 11 percent (USDI 2006b, 25), and the second-highest participation rate in wildlife-watching activities (USDI 2006a, 112).[9] Vermont is also over 97 percent white (U.S. Census Bureau 2000a, 28), which is significant in light of the fact that 96 percent of American hunters are white (USDI 2002, 31). Addison County mirrors these state-level trends, but with 35,974 residents and a population density of 46.7 people per square mile at the time of the 2000 U.S. Census, it is slightly more rural than average for Vermont (U.S. Census Bureau 2000b). Middlebury is the largest town in Addison County, with a population of 8,183. There are twenty-one other towns in the county, only three of which have more than 2,000 residents: Bristol (3,788), Vergennes (2,741; it is officially designated a city), and Ferrisburgh (2,657). Nearly 87,000 acres of the county lies within the Green Mountain National Forest, including the smaller towns of Lincoln (1,270) and Ripton (556).

Addison County is much more than a *demographically* rural county.[10] It is a place where traditional rural practices like hunting are not simply tolerated but enthusiastically pursued and publicly celebrated. For example, during the fall rifle deer season the local newspaper, the *Addison Independent*, lists all the people who have killed deer along with each animal's size and sex. Wild game dinners in Middlebury attract hundreds of people, and there are still well-attended opening-day breakfasts held in the pre-dawn hours of the first day of deer season.

One important characteristically "rural" aspect of life in rural Vermont is the prominence of *localism*: a preference for one's local community based in the perceived uniqueness of that community.[11] Indeed, newcomers and visitors to Vermont are often immediately struck by the strong insider/outsider distinction that permeates everyday life.[12] Vermont residents are well aware of the heightened social status that comes with being a native and the stigma that comes with being pegged as a "flatlander" (a derogatory term used by Vermonters to refer to out-of-staters).[13] And even though many of Vermont's most well-known residents are flatlanders (such as the state's former governor, Howard Dean, and U.S. senator Bernie Sanders, as well as the artist Woody Jackson and the founders of Ben & Jerry's Ice Cream), transplants to Vermont are still surprisingly defensive about their non-native status. For these people, it is especially anxiety-provoking to be asked where they grew up. After much foot-shuffling, the guilty party will sheepishly admit that, yes, they have only lived in Vermont for the past twenty-five years.[14]

On several occasions, I was reminded of the extent to which a deep attachment to the local physical environment contributes to rural Vermonters' localism. Perhaps the best example of this came from a man in his early twenties at a deer camp in Ripton. Several of us were chatting about what it would be like to hunt in Wyoming or Colorado, and this young man said that he just wasn't interested in hunting anywhere except the places he'd always hunted growing up. And he meant it. Even other places in the state of Vermont were too far-flung to lure him away from the mountains close to his home. While I wasn't surprised by this man's love of his local terrain, I was surprised when he named a place in the mountains no more than a couple of miles from the deer camp where we were talking as an example of a distant locale that he wouldn't have any interest in visiting during the hunting season. To be sure, for many of the rural Vermonters that I have met over the years, Vermont is truly their "homeland."[15]

In large part because of the presence of Middlebury College, both the educational levels and the incomes of Addison County residents are somewhat higher than in many other rural American counties.[16] For example, 29.8 percent of Addison County residents twenty-five or older have bachelor's degrees, and 11.9 percent have graduate or professional degrees. At the time of the 2000 census the median household income of Addison County was $43,142, compared to $40,856 for Vermont and $41,994 for the United States as a whole; only 5.1 percent of Addison County families were below the poverty level, again a more favorable percentage than both the Vermont median of 6.3 percent and the U.S. median of 9.2 percent (all figures from U.S. Census Bureau 2000b).[17]

The primary reason for this economic health is the availability of a wide range of jobs. Though it does not offer a particularly vibrant retail community,[18] Addison County is home to a rather impressive assortment of small manufacturers that produce a considerable array of products, ranging from custom clothing and homemade soap to paper office supplies and plastic injection molds. Some of the better known manufacturers are Danforth Pewterers in Middlebury, Geiger of Austria (wool clothing) in Middlebury, Cabot (dairy products) in Middlebury, Otter Creek Brewing (a microbrewery) in Middlebury, Autumn Harp (cosmetics) in Bristol, and Goodrich Corporation (aerospace components) in Vergennes. And 6.9 percent of Addison County residents are employed in the industrial category referred to by the U.S. Census Bureau as "Agriculture, forestry, fishing and hunting, and mining." This far exceeds both the overall Vermont (3 percent) and national (1.9 percent) employment levels in these industries. The great majority of these jobs are in agriculture and forestry, testifying to the continued existence of "traditional" rural practices in Addison County (U.S. Census Bureau 2000b). Bordering counties—particularly Rutland to the south and Chittenden to the north—also provide many jobs for Addison County residents.

The data in this book was gathered in a variety of ways over more than ten years.[19] Participant-observation (the anthropological term for taking part in everyday activities as a way of learning about a particular group of people) has been a key component of my research agenda, as it is for most cultural anthropologists. Because I held two part-time jobs during my fieldwork (as a clerk in a local sporting goods store and as a waiter/bartender at a local restaurant) I had ample opportunity to interact with and observe a variety of people in a wide range of situations. I also spent time at nine Green Mountain deer camps, where I participated in all areas of camp life, from hauling water, washing dishes, and playing cards to climbing in a pickup truck at the break of dawn with a high-powered rifle in my hands in search of white-tail bucks. My research at deer camps continues, and I've made multiple return trips to some of the original nine. More recently, I've conducted additional research at a controversial Addison County coyote-hunting tournament.

The second main source of my data has been open-ended ethnographic interviews. During my initial research I conducted fifty formal interviews, twenty-eight with men and twenty-two with women),[20] as well as countless informal interviews at times when, for whatever reason, the situation wasn't right for sitting down for a more structured interview. Since the completion of my initial fieldwork, I have continued to interview Vermont hunters, but I have spent more time talking with hunting opponents and wildlife management officials than I have in the past.

Considering American hunting's intriguing combination of violence and extreme gender specificity, it is truly surprising that so little attention has been paid to this topic by the anthropological community. Only one ethnographic study focusing on Euro-American hunting has been published in the history of cultural anthropology, Stuart Marks's *Southern Hunting in Black and White* (1991).[21] Yet the field of cultural anthropology itself is well suited for such an endeavor because one of its long-held priorities has been to understand how particular groups of people think about their physical environment—a process sometimes referred to as the socialization of nature (Descola 1994). Such an approach, by definition, entails the understanding that "nature" is a cultural construction rather than a universal physical entity that is thought about in exactly the same way by every group of people the world over.[22]

Since the early 1990s, several nonacademic books have been written about various hunting-related issues, including wildlife ecology, wilderness philosophy, and the ethics of hunting. Their authors strive to emphasize the more contemplative and intimate aspects of this often controversial activity (Nelson 1997; Petersen 1996; Swan 1995; Posewitz 1994; Kerasote 1993; Sajna 1990). A common theme of these works is the characterization of hunting as a respectful, emotional, and perhaps even essential engagement with the nonhuman world. Whether examining former president Jimmy Carter's childhood in rural Georgia (Petersen 1996, 35–47) or commenting provocatively on the relationship of environmental degradation and vegetarianism (Kerasote 1993), all these authors attempt to move the contemporary discussion of hunting beyond stereotypical themes of masculine domination and unbridled human malevolence toward the nonhuman world. I, too, will discuss aspects of contemporary Euro-American hunting practices that resist conventional stereotypes. My conclusions are made at the *cultural* level, however, and they emanate from systematic ethnographic research that seeks to locate consistent patterns of meaning associated with Euro-American hunting in rural Vermont.[23]

The great majority of academic work on hunting in America has been undertaken by researchers operating under the rubric of sociology, or at least by those emphasizing quantitative research methods rather than qualitative ones. These scholars have succeeded in producing the accepted "profile" of the American hunter. As a result, we now know, for example, that hunters are generally white, rural, male, middle-income manual laborers whose fathers are hunters (Heberlein and Thomson 1991; USDI 2002). This work has also identified important social trends in the social production of male versus female hunters (Heberlein 1987); and some have speculated about the different stages hunters pass through as they age (Jackson and Norton 1980).[24]

Historians have also made a significant contribution to the study of Euro-American hunting, in works such as Bill Cronon's *Changes in the Land*, Richard Slotkin's *Regeneration through Violence*, Louis Warren's *The Hunter's Game*, Karl Jacoby's *Crimes against Nature*, and Daniel Herman's *Hunting and the American Imagination*. These books remind us that for Euro-Americans cultural politics and moral ambivalence have always ridden shotgun with the topic of hunting, from the earliest colonial days when agriculturally minded Puritans morally condemned American Indians for their hunting ways.

Perhaps the greatest cost of the dearth of qualitative work on hunting is the effect (or lack thereof) on public discourse involving hunting. As the debate surrounding hunting becomes increasingly contentious, the need for ethnographic, meaning-centered studies becomes more pressing. An ambivalent, nonhunting public should know not only who hunts, where these people live, how much money they earn, and their level of education, but also *what hunting means to hunters*. The rural sociologist Thomas

Heberlein wrote in his foreword to Robert Wegner's *Legendary Deer Camps*: "Although I am a sociologist, I hunt deer during the season and have not snooped around many camps in the name of science. This would be a good project for a young anthropologist, I think" (Wegner 2001, 7). It is my hope that this book will take an important step in this direction.

1

From Extinction to Tradition: Wildlife Management

No understanding of local hunting is possible without some reference to the state's institutions and their power over the lives of ordinary citizens.

STUART MARKS, *Southern Hunting in Black and White* (1991)

If there's one single subject on which every single Vermonter feels he or she is an expert, it is management of the deer herd.

Brattleboro Reformer, November 30, 1978

In his richly textured analysis of hunting in rural Scotland County, North Carolina (*Southern Hunting in Black and White*), Stuart Marks discusses the varying amounts of influence that different constituencies of hunters have over game management regulations. This is a precious kind of influence to wield since, as Marks notes, "state laws circumscribe the boundaries within which game is taken" (1991, 65). The ways in which animals are taken, in turn, have a considerable influence on the kinds of experiences people have while hunting and the specific meanings that will emerge from those experiences. In the case of Scotland County, Marks describes a stratified (particularly along racial and economic lines) hunting community in which certain hunters (whether through wealth, social connections, organizational skill, or simply a deep and consistent engagement with the public aspects of wildlife management) have the ability to influence policy decisions. "Yet for many others," Marks writes, "such regulatory protocols remain mysteries. For members of this latter group, hunting remains largely a traditional activity, local in content and context. Their prospects for transmitting their passions and traditions to succeeding generations seem as problematic as their access to surviving game" (65).

Vermont offers an interesting contrast to the situation Marks describes. Whereas in North Carolina "local" practices and perspectives seem to occupy a marginal political position, in Vermont the opinions of "locals" are still a powerful force in state affairs.[1] Politicians are keenly attuned to local issues and will pay dearly at election time if they choose to ignore them. It should come as no surprise, then, that "regular people" still have a strong and influential voice in the process. By contributing to wildlife management discussions, rural Vermonters are assuring that control over one of the most important ingredients in the production (and reproduction) of local rural Vermont culture—hunting—is not completely relinquished to the state. Indeed, it is hard to underestimate the importance of hunting to local identity for some rural Vermonters. I well remember when one hunter said to me, "Every Vermonter deserves a chance to get a deer." I knew hunting was woven deeply into the fabric of rural Vermont; nevertheless, I was shocked by the extent to which this man was combining deer-hunting and state identity. His comment takes on even more ethnographic significance when one considers just how important local identity is to so many Vermonters. It also seems to imply that Vermonter's have "always" hunted deer. As we will see in this chapter, however, the legendary Vermont deer-hunting tradition is actually of relatively recent vintage.

Vermont Game Laws

The creation of wildlife management regulations in Vermont is a complex political process that involves three separate governmental bodies: the Vermont State Legislature, the state's Fish and Wildlife Board, and its Department of Fish and Wildlife (called the Department of Fish and Game until 1983).[2] Unlike many other state legislatures, Vermont's legislature possesses broad powers to create specific wildlife management laws. The Fish and Wildlife Board consists of fourteen members (one from each county) appointed by the governor; it is charged with adopting rules "for the regulation of fish and wild game and the taking thereof except as otherwise specifically provided by law"; the Department of Fish and Wildlife conducts scientific research and makes management recommendations to the Fish and Wildlife Board (Vermont Statutes, Title 10, chap. 103, sec. 4082). The board also takes into account citizen testimony at public hearings.

In short, the board—a panel of political appointees—creates game management regulations based on the recommendations of professional biologists employed by the Department of Fish and Wildlife—that is, unless the legislature decides to take over the process. In that case, neither the board nor the department has any official control over the outcome, because their powers are granted by the legislature and can be overridden by it at any time.[3]

A prime example of the legislature's overarching authority, as explained to me by John Hall, information director for the Fish and Wildlife Department, involves the regulation of the hunting of northern pike with firearms in Lake Champlain. In 1969 the department proposed to ban this practice; the board subsequently passed this new regulation. The legislature disagreed, however, and promptly wrote a statute declaring such hunting legal until decided otherwise by a future legislature.[4] Here the Vermont Legislature acted to preserve and protect a traditional rural practice against the recommendation of professional game biologists.

Even those involved with crafting Vermont's game management regulations are sometimes astounded by the complexity of the process. In a 2002 e-mail to the current chair of the Fish and Wildlife Board, Rob Borowske, I asked him to clarify the distribution of power among the board, the Fish and Wildlife Department, and the legislature. "GREAT question," he responded; "I sometimes struggle with that one also. It is not always clear who has what power or authority. . . . The Legislature establishes the laws or statutes. But it can be confusing: for instance, the Board has the authority to manage the health of the deer herd by creating rules and regulations of the hunt and also establish yearly the number of antlerless deer to be taken. However, the legislative Committee has the authority to state what a legal deer is, and when the seasons are."[5]

Not surprisingly, the peculiar distribution of powers over the creation of game management laws in Vermont is often viewed by outside observers with considerable skepticism. But from an anthropological perspective, the relationship shared by the state legislature, the Fish and Wildlife Board, and the Department of Fish and Wildlife—and the laws and regulations that emerge from their interactions—presents a fascinating study of rationality and science, local human–nature relations, state identity, and the articulation of state power.

The Judicial Roots of Vermont Game Management

Although the Vermont legislature possesses an unusual degree of power over the creation of game management legislation, generally speaking Vermont's game management is quite similar to every other state's. The judicial roots of American wildlife management are found in English common law.[6] Like Roman common law before it, English common law, established in the twelfth century, holds that wild animals are "common property" until either killed or captured, at which time they become private property (Tober 1981, 23; Shaw 1985, 4). This did not mean, however, that game was originally viewed as the property of the people, as James Shaw notes: "The king and his nobles claimed formal legal ownership of game and exclusive hunting rights until the signing of the Magna Charta in A.D. 1215. That document transferred ownership of wildlife from the crown to the people. Since then wildlife has remained public property tended by the government as a 'sacred trust'" (Shaw 1985, 5).[7]

The way people actually interacted with wild animals in the fledgling American colonies was dramatically different than what was occurring at the same time in England. There, regardless of the dictates of common law, rigid land-use restrictions reserved hunting for the elite. Not surprisingly, this legacy of aristocratic privilege in England made equal access to game a primary concern of colonial Americans. In the words of Robert Muth and Wesley Jamison, "The concept of 'the king's deer' and the English legacy of private game preserves exclusively reserved for the hunting pleasures of the nobility were incompatible with the democratic precepts of the American Revolution" (2000, 842). The result in the colonies and states was a democratized approach to wildlife that amounted to an important expression of the expanded freedoms offered in the New World (Sherwood 1981, 20; McCandless 1985, 13). As James Tober explains: "Wildlife was there for the taking. It was a common heritage, not subject to restrictive controls which smacked of Old World class structure" (1981, 17). Vermont took these "restrictive controls" head-on when, in 1777, it became the first state to constitutionally guarantee one's right to hunt and fish (chap. 2, sec. 67). This law also provided wide land-use rights to *private property*, and to this day hunters are free to hunt on private property unless they are restricted from doing so by landowners. It should come as no surprise that native/flatlander tensions often flare up around this issue, since newcomers to the state (particularly if they don't hunt) are often uncomfortable with the idea of strangers hunting on their land.

Over the course of the colonial period, Euro-American settlers all across what is now the United States annihilated wildlife populations through unregulated market and subsistence hunting and, most importantly, habitat destruction. This onslaught is often characterized as the inevitable outcome of the combination of seemingly inexhaustible wild animal populations and colonists intent on amending the class-based access to land and game that was so detested in England (Muth and Jamison 2000, 843; Gilbert and Dodds 1987, 6; McCandless 1985, 13; Tober 1981, 17, 27). Before long, in an effort to slow the decline of animal populations, hunting regulations were enacted in colonial America. By the time of the American Revolution in 1776, twelve of the thirteen American colonies had hunting restrictions in place.

Not deviating from this broader colonial trend, animals and their habitat were eliminated at an astonishing rate in the Republic of Vermont and, later, the state of Vermont. For example, the portion of Vermont covered by forest is estimated to have declined from 95 percent in the early seventeenth century to 25 percent in the mid-nineteenth century (Klyza and Trombulak 1999, 67, 93).[8] The destruction of these forests, in conjunction with unregulated hunting, led to the decline of many wild animal populations: "Roughly one hundred years after European settlers began to stream into Vermont, the state was a landscape transformed. The native peoples were largely gone, old-growth forests had virtually disappeared, large animals such as the elk, the mountain lion, the wolf, and the wolverine were gone or virtually so" (ibid., 85–86).[9]

Deer Management

Perhaps the most anxiety-provoking colonial wildlife population decline occurred among white-tailed deer. Even today, no other wildlife management issue comes close to eliciting the emotion of deer-hunting regulations. Then again, no other form of hunting is pursued as passionately or plays such an important role in local culture.[10] As a result, an examination of the management of deer provides important insight into how local culture, legislative action, and wildlife management science coalesce in the creation of both hunting regulations and a celebrated tradition.

The decimation of deer herds was a common theme in colonial New England. In 1646 Rhode Island established a closed season on deer. Connecticut (1698), Massachusetts (1694), New York (1705), and New Hampshire (1741) followed with their own game management regulations. Then, as Leonard Foote explains: "In February, 1779, one of the first acts of the newly formed Vermont legislature was to prohibit the taking of deer from January 10th to June 10th. A penalty of fifteen pounds was provided, and the town Justice of Peace was 'to hear and determine.' At the same legislature, a penalty of thirty shillings was imposed for shipment of hides which were 'raw or untanned' out of the state" (1944, 16). Finally, in 1865, the deer season was closed indefinitely in an attempt to preserve Vermont's few remaining deer, which were mostly living in Essex County in the far northeastern reaches of the state (Titcomb 1898, 61). Even so, a robust recovery of the deer population did not begin until April 1878, when seventeen deer from New York were released in southern Vermont and the legislature enacted a moratorium on deer hunting that would last nearly twenty years. The small herd grew rapidly, and by the mid-1890s farmers were complaining bitterly about deer-related crop damage and pressuring the legislature to reopen the deer season.

While the reintroduction of deer was certainly critical to the veritable explosion of the Vermont white-tailed deer population that would continue through the mid-twentieth century, without some other overarching social and ecological transformations that were simultaneously occurring in Vermont the herd would not have rebounded as dramatically as it did. First, by the conclusion of the Civil War, Vermont's agricultural economy was in trouble and many farmers were leaving the state.[11] Abandoned and overgrown farms, combined with a dearth of large carnivores, provided important habitat for the fledgling herd (Klyza and Trombulak 1999, 166; Pyne 1996, 16; Tober 1981, 70). Second, by this time competition from western sheep ranchers had turned Vermont from a sheep state to a dairy state.[12] This transition had an important effect on the Vermont landscape because unlike sheep, which could survive on the low-quality browse found on the marginal land that resulted from the state's timber industry, dairy cows depended on the rich pastures of the valleys, thereby allowing forests to regenerate, deer to flourish, and a hunting tradition to take root in the rocky soil of the Green Mountains.[13]

In 1896, the legislature responded to citizen pressure and passed legislation that created an open deer season for the entire the month of October. On October 1, 1897, the first state-wide hunting season (bucks only, no hunting dogs) commenced. Despite enthusiastic (and highly exaggerated) press reports that attracted many out-of-state hunters, only 103 bucks were killed; possibly another fifty were taken illegally (Perry 1964, 36; Tober 1981, 70). By 1899, the season was reduced to ten days and only ninety bucks were taken. In an article written the following year, the state's fish and game commissioner, John W. Titcomb, offered a sardonic commentary on the events that led up to the reopening of the deer season: "For many years past, reports in the city papers, notably in the Sunday editions, have exaggerated the conditions in Vermont as to the abundance of deer.

The advent of an open season was the signal for more frequent articles with proportionately increased exaggeration if such is possible. The unsophisticated were led to believe that the ravages of deer resembled the grass-hopper plague of the west, and that it would be an easy matter to shoot a deer on the first day of the open season" (1898, 61).

Toward the conclusion of a discussion of late nineteenth-century deer management, Titcomb presents a not-so-veiled critique of the legislature's decision to open the deer season: "The press of the State very generally condemns the idea of an open season, and is supported by the real sportsmen. Our forests are not extensive enough to warrant having much of an open season. The writer believes Vermont is more attractive with a few live deer, which can be occasionally seen when pleasure driving, than with an open season and consequent slaughter of half-tamed animals" (1898, 63).

Nevertheless, with the dawn of the twentieth century, concerns about the size of the Vermont deer population continued to mount, particularly among farmers, who again pointed to heavy crop damage. Now, less than thirty years after deer were reintroduced, many considered them to be overpopulated: "Deer repellants were experimented with. Sulphur was tried and seemed efficient for only a short time; then a kerosene emulsion was tried which proved to be more effective in eliminating the garden crops than the deer" (Perry 1964, 37). In response to the outcry over crop damage, the legislature allowed the harvesting of antlerless deer during the 1909 hunting season—the first of five "either sex" seasons between 1909 and 1920 (Pyne 1996, 16; Perry 1964, 37–38).

As we will see, antlerless seasons, sometimes called doe seasons, are the method of choice for thinning the deer herd in Vermont (as is the case for the United States generally).[14] The purpose of an antlerless season is to reduce the number of reproducing females in the herd, thereby helping to reduce the overall size of the herd. Male deer without antlers, or with smaller antlers than would normally be allowed in the regular deer hunting season, can be legally killed during this season as well. The fact that antlerless seasons are instituted by state game management agencies when deer populations are deemed to be to unhealthily large (thus proving that hunting bucks is not the most efficient means of populations control) does little to allay the common nonhunter suspicion that "wildlife management" is really nothing more than "target management" for hunters. For many critics of hunting, this apparent contradiction in hunting motivations and consequences seriously undermines the ethical legitimacy of hunting.

The Vermont deer herd gradually increased after 1920, and by 1940 the buck kill (the number of bucks killed during the deer hunting season, an indication of the overall size and health of the herd) exceeded 3,000 for the first time. In 1950 the buck kill topped 6,000, and in 1957 over 11,000 deer were killed during a sixteen-day rifle season. By 1960, the fall deer season was "a Vermont tradition chiseled in granite. . . . As the herd peaked at about 250,000 in the mid-1960s, Vermont offered some of the best deer hunting in the country" (Pyne 1996, 17). In 1965, with a buck kill that exceeded 16,000 for the first time, the health of the Vermont deer herd probably seemed to good to be true. Indeed, among those old enough to remember, the 1950s and 1960s are spoken of as a golden age of deer hunting. As longtime hunter Ralph Larrow told Lawrence Pyne, "Oh, it was something back then, you never saw so many deer" (Pyne 1996, 17).

In retrospect, all those bucks being taken from a relatively small state probably was, in fact, too good to be true. In a 1947 Vermont Fish and Game Service publication titled *The Time Is Now!—A Pictorial Story of Vermont's Deer Herd*, wildlife biologists passionately argued that the state's deer herd was standing on the precipice of imminent disaster as a result of overpopulation. The simple argument—made with narrative, charts, and photos of dead, frozen deer—was that the Vermont deer herd was exceeding the maximum carrying capacity of the land and that "one or more open seasons on antlerless deer" were necessary to regain the balance between the deer herd and "available winter range" (Seamans 1947, 3, 35). "Vermont is now on the threshold of a period during which we can act on the experiences of others or blindly permit unnecessary losses in our deer herd. Already too much time, energy and money has been spent to bring the herd to its present position to allow this to happen" (3).

All the while, however, it is clear that the authors were fully cognizant of the fact that their biggest challenge was not to prove their case to other biologists, but rather to make a case to ordinary Vermont hunters about the legitimacy of scientific wildlife management: "The conditions as they exist, the prediction of what will come and the suggestions that this publication is attempting to set forth, may be interpreted by some hunters as being the death knell to his sport. The truth is a far cry from any such gloomy pictures. Management is the application of scientific study for the utilization of natural resources. Game is one of the most important resources we have and it must be maintained for all generations to come" (Seamans 1947, 22).

The Time Is Now! is trying to prove not only that scientific wildlife management works, but also that it can be *trusted*. Recall that earlier attempts at antlerless seasons between 1909 and 1920 actually killed too many deer, and the historical legacy of those management decisions were working against the 1947 managers. The authors attempted to alleviate such fears: "A need for a controlled antlerless deer season is demonstrated by considering the open seasons of 1909, 1910, 1915, 1919 and 1920. More does and fawns were shot than the herd could stand, resulting in lower kills for a number of years. The method of reducing the population, as set forth in this bulletin, will prevent this from happening again" (ibid., 23) The suggestions of the wildlife managers were not immediately followed, but the legislature did eventually respond by setting one-day antlerless seasons in 1961 and 1962. Apparently, this measure had little effect on the overall herd density (Perry 1964, 43).

In 1966, for the first time, the state legislature transferred its deer management responsibilities to the Fish and Game Board. Acting on the concerns of state biologists, the board approved four antlerless seasons over the next five years. As fate would have it, the implementation of these antlerless seasons coincided with the notoriously harsh winters of 1969–70 and 1970–71, in which tens of thousands of deer fell victim to exposure and malnutrition. By 1971 the buck kill was only 7,760 (down from over 17,000 in 1966), the lowest it had been in nearly twenty years. When discussing deer management it is common for hunters to reflect on the winters of 1969–71, describing in graphic detail the deer carcasses that were littering deer yards in the mountains. Significantly, these are also the kinds of testimonials that seem to be the most effective in arguing that killing individual deer can actually be thought of as an act of compassion toward the overall herd.

Of course, this staggering loss of deer was just the sort of calamity that deer biologists had anticipated. Many hunters, however, disagreed with the biologists' scientific explanation for the sudden decline in the deer population and placed the blame for it squarely on the antlerless seasons. This correlation of a devastating winter die-off of deer and the antlerless seasons only confirmed people's worst suspicions about the wisdom of killing does. Even today, many Vermonters are adamantly opposed to shooting does. Generally, their opposition to this practice is based on one of two beliefs: that does are female and thus should be spared (for example, I have heard does referred to as "life-givers," or "the nurturing side of nature"), or that too many does will die when the hunting season deaths are combined with the "winter kill"—the animals that starve to death over the winter. Another explanation I have been given for the lack of support for antlerless seasons is that hunters know that many does are killed illegally each year for "camp meat" (venison to eat at deer camp during the hunting season) and feel that any more killing of does would be detrimental to the health of the herd.[15]

An acute sensitivity to the topics of weather and state management regulations has dominated my conversations with hunters about the health of the deer herd. These, of course, are perfectly reasonable topics to discuss. What has always surprised me, however, is the extent to which habitat *does not* come up in these conversations. Indeed, the hunter who describes habitat as a key determinant of the health of the deer herd is a truly noteworthy informant. On the other hand, in my conversations with Fish and Wildlife personnel, habitat "conversion and fragmentation" are considered the most important factors influencing the deer herd. (The Department of Fish and Wildlife even tried to hammer home the importance of habitat by distributing "Habitat Is the Key" key chains in the early 1980s. The key chains will soon be reissued as part of a "Welcome Wagon" kit that the department will provide for landowners who purchase more than twenty-five acres.) Considering how much so many hunters know about the outdoors, I must admit I was surprised by the collective blind spot on habitat. I falsely assumed that people in a rural area with both farming and an active timber industry would be attuned to the importance of an abundance of good habitat to the health of the deer population. To be sure, there are people who actively plant certain crops (such as soybeans) to attract and feed deer, but the more common conversation I have had is with frustrated hunters trying to figure out "why there aren't any deer in Vermont" and usually blaming the Department of Fish and Wildlife. Among the hunters I have encountered in Vermont, ecological knowledge is usually specific and suited for the situations that a particular person encounters most often, rather than a systematic knowledge of the ecology of an entire region

Perhaps the importance of habitat goes unnoticed because, in the region where I do most of my research, the "problem" with habitat isn't that it's going away—much of it is contained within the Green Mountain National Forest—but that much of the forest is uniformly old, and therefore lacking sufficient browse for deer. Once again, the fact that so few hunters mention this when talking about the deer herd is truly surprising, since many of them intentionally hunt in clear-cuts because they know deer seek out such areas to feed on young vegetation.

Heeding their rural constituents, the legislature reclaimed managerial authority of the deer herd in 1971 and swiftly implemented an eight-year moratorium on doe hunting (Pyne 1996, 17). But when this plan failed to grow the deer herd to the public's satisfaction (at 7,087, the 1978 buck kill was the lowest it had been since 1952), in 1979 the legislature again transferred control of the deer herd back to the Fish and Game Board. As Lawrence Pyne aptly notes, by this time deer management in Vermont had become a "political hot potato" (1996, 17). Making the most of its newfound powers, the board moved decisively to institute more antlerless seasons. There would be eight over the next nine years, and they would account for the taking of 55,790 antlerless deer. The all-important buck kill, however, was not rebounding as quickly as hunters desired, and for the better part of a decade the doe seasons were frequently debated; especially since in 1987 (the last of the eight antlerless seasons that were initiated in 1979) the buck kill was only 5,903. Consequently, in 1991 the legislature prohibited all hunting of antlerless deer during the rifle season.[16] On the other hand, also in 1991, hunters, legislators, and biologists finally arrived at a compromise position that seems to be working: antlerless seasons are now held during the shorter and less popular muzzle-loader deer hunting season that follows the regular November rifle season.

The advent of the muzzle-loader antlerless season is considered a watershed event in Vermont deer management because it allows biologists to have a reasonable amount of control over the herd. As John Hall of the Department of Fish and Wildlife explained to me, the department felt an antlerless rifle season was critical to effective deer management and was always reluctant to compromise on this. Hunters were not convinced that it was a good idea, however; according to Hall, they were afraid that too many does would be killed by hunters disobeying the hunting laws. For their part, the legislators felt that Fish and Wildlife was not sufficiently listening to hunters' concerns. The net result, Hall said, was that "we were basically powerless to manage the deer herd." Finally, the compromise was reached after Fish and Wildlife decided that it was "more important to manage the herd effectively, rather than give it up for eight years [a reference to the 1971 moratorium on antlerless seasons]" (Hall, personal communication). Today, in addition to the muzzle-loader season, the Youth Day Deer Hunting Season (which began in 2001 as an

explicit attempt to recruit youngsters into hunting) serves as an additional
antlerless season.

To anyone who has not been involved in discussions of deer management in Vermont (which might well be similar in tone to discussions in
certain areas of such states as Pennsylvania and Wisconsin), referring to
deer management as "a political hot potato" might seem a bit dramatic.
It's not, as these vivid passages from an article by Edward Cronin in a 1979
issue of *Country Journal* show:

> At most, I assumed it would be something like our regular town meeting.
> The auditorium would be loosely filled with interested but quiet spectators.
> The meeting would be dominated by a vocal few who spoke out strongly
> about their concerns. There would be vigorous debate on some issues,
> but, in general, each vote would be duly recorded and the town moderator
> would maintain the decorum of the proceedings.
>
> Instead, as I approached our local grade school I saw row after row of
> cars lining the streets and parked at every odd angle as if abandoned in
> an emergency. Suddenly, I was caught up in a flow of anxious people that
> swept me inside; the quiet of our snowy landscape was exchanged for a
> smoke-filled, noisy auditorium. The room was packed tight. The men, clus-
> tered here and there in small groups, talked loudly among themselves and
> gestured back and forth as if arguing a life and death matter. The mixed
> conversations produced a roar that made the room seem small, and there
> was a tension in the air, an energy that swayed over the crowd like waves
> on an angry sea.
>
> What issue could possibly draw such a gathering and bring forth such
> unbridled emotions? The Equal Rights Amendment? Abortion? The path
> of a new interstate? No this was Vermont. The question at hand was the
> management of the deer herd. (1979, 100)

The Vermont deer herd gradually rebounded from the anxiety-ridden
days of the mid-1980s, and by 1993 the buck kill reached 10,000. It continued at that level until 2001, when, after a harsh winter, it dropped by
nearly 3,000.[17] In 2004 the buck kill (5,589) was the lowest it had been since
1982, and it prompted what *Rutland Herald* outdoor writer Dennis Jensen
calls "the most revolutionary changes in deer hunting in Vermont in 100
years": the 2005 implementation of a ban on shooting spiked bucks (bucks
with only two single antlers, often called spikehorns) during the November firearms hunting season (Jensen 2007, 1). So far, the revolution appears to be working according to plan. After an initial decline in the buck
kill, because about 33 percent of the bucks taken annually were usually
spikehorns, it has rebounded strongly. During the 2008 season the total
deer harvest was 17,046 and the buck harvest was 9,539 (VFWD 2008, 4).

Not only is this mangement scheme increasing opportunities to kill deer in general, but it has also increased the chances of seeing older, and therefore larger, deer.

Transitions in Human–Animal Relationships

As I will discuss in great detail in chapter 2, "stewardship" was a central theme in my conversations with hunters, and to them an important aspect of this stewardship was the idea of killing animals, including deer, for the sake of the overall health of a particular species. For a long time I accepted these testimonies at face value, assuming that a stewardship ethic that involved keeping populations in check was a long-held practice among Vermont hunters. After all, the state of Vermont had seen its fair share of environmental transformations over the years, and many of the people I talked to had some connection with the stewardship of domestic livestock. But as time passed and I tried to make sense of my experiences in Vermont, I began to wonder to what extent this idea of killing deer for the sake of the herd might actually be a recently invented tradition. My first thought was that it might be the product of pro-hunting marketing efforts, since I knew that many hunters, in response to antihunting critiques, were quick to justify hunting as a biological necessity. In any event, I made some long-distance phone calls and asked a couple of people in Addison County what they thought.

The two people with whom I spoke—one woman, one man—are both over the age of fifty; both are natives of Vermont and have a deep knowledge of local history. And both of them confirmed my suspicion—to a degree. The woman, an avid deer hunter, said that all this talk about hunting as a way of maintaining healthy animal populations was clearly not something that she had grown up hearing: "When I was a kid I don't ever remember an overpopulation of anything." When she was young, she said, people were more concerned with getting something to eat than with population densities and doe seasons.

The man, a retired high school biology teacher and member of a local deer camp, had similar memories: "I don't ever remember discussions about herd health in those times." A little later he added, "I think we were educated by the game biologists." He recalled the beginnings of game management in New England and specified the 1970s as the time that it was introduced in states such as Vermont, New Hampshire, and Maine. The main conflict that he remembered was between biologists who were encouraging people to hunt for does in an effort to reduce herd growth and locals who were often strongly opposed to the idea. The "old-timers," as he put it, contended that weather should be the mechanism by which deer herds are thinned.

Considering the history of deer management in Vermont, it is easy to understand why these two hunters and many of the other (mostly middle-aged) hunters I have met think of deer management as something that commenced in the late 1960s the 1970s. After all, during their childhoods the herd did nothing but grow, and controversial ideas like doe hunting were not often discussed. What they saw as the beginning of deer management was actually a dramatic shift in the role of hunting in the overall deer management scheme. By the 1970s, increasing the size of the deer herd was no longer the unquestioned goal of deer management. Instead, game managers were attempting to maintain the health and size of an existing herd. As the rural sociologist Thomas Heberlein has explained, this shift in deer management (which has now occurred all over the United States) marks a complete reversal of Aldo Leopold's classic definition of game management as "the art of making land produce sustained annual crops of wild game for recreational use" (Leopold 1947, 3) Instead, hunting is now a *mechanism* by which game managers keep deer populations at healthy and socially acceptable levels (Heberlein 2002).[18]

The changes in perspectives on deer management that have occurred in Vermont since the 1970s were echoed in my conversation with John Hall. He began his tenure with the Department of Fish and Game (as it was then called) in the late 1960s, and one of his first assignments was to head off into the Green Mountains and make a 16mm film of the thousands of deer that were starving to death during the infamous winters of 1969–70 and 1970–71. In 1971 he began showing the film in public schools across the state in an effort to convince young people of the necessity of scientific deer management techniques—namely antlerless seasons.

At the time, of course, many Vermonters were vehemently opposed to the idea of doe seasons as the prime means of deer population control. Thirty years later, however, Hall notes that he has witnessed a great shift in opinions about killing does for the sake of the general health of the herd: "Looking at today's hunters . . . shooting does is not what they want to talk about." What do they want to talk about? According to Hall, the answer to that question is QDM, or *quality deer management*. He explained that people in their twenties and thirties generally have no qualms about killing does if it is part of a general management plan that will provide

Vermont's deer hunters with more opportunities to hunt trophy deer. Obviously, the positive reception of the current moratorium on spikehorns is testament to attractiveness of QDM.

Quality deer management marks a historic coalescence of the deer management philosophies of rural hunters and Fish and Wildlife biologists. As the people who were in their forties, fifties, and sixties during my initial fieldwork in the late 1990s become a smaller fraction of Vermont's population, resistance to the idea of killing does to manage the deer herd will almost certainly become a distant memory. This also means that it is highly unlikely that any future Vermont legislature will enact increased restrictions on doe hunting in an attempt to speak for "the people" of Vermont.

While QDM may be an unqualified success for deer management in Vermont, it could well alter local meanings of hunting because of its emphasis on "the trophy." Most hunters, while certainly not opposed to the idea of bagging a "Rackasauras" on opening day, are thrilled to bring home any deer at all. If it happens to be unusually large, or has a trophy rack, so much the better. But QDM is a different philosophy. It focuses on the size of deer and/or its rack as a way of determining the value of a hunting experience. This thought first crossed my mind when I initially learned about QDM in 2002 or 2003, and was it emphasized again in a conversation with a Vermont game warden in 2004, who said, "The cultural perception of hunting has gone from process to product. . . . They're taking the hunt out of hunting." A man at a local deer camp shared similar sentiments and pointed out (even though he agreed that it might be good for growing bigger deer) that QDM was a completely different approach from what he referred to as the Vermont "family" hunting tradition, which is not oriented around a quest for trophy bucks but rather around the love of fresh venison, the enjoyment of family and friends, and the chance to spend some time in the woods rather than at work. Considering how many times I have heard hunters say "You can't eat the horns," I think this guy had a point.

Since the beginning of state-sanctioned wildlife management in Vermont in 1779, when the creation of wildlife management regulations was solely the job of the legislature, the process has become infinitely more complex. Today, wildlife management laws are created only after long and sometimes heated negotiations between elected officials, political appointees, and professional wildlife biologists. These negotiations often involve considerable disagreement between legislators representing the "local culture" of rural constituents and Fish and Wildlife biologists operating from the perspective of scientific wildlife management.

While one might expect the Western rationality of the biologists to prevail in these confrontations, in Vermont the state legislature is empowered to override the recommendations of the Department of Fish and Wildlife. As a result, the state's wildlife management regulations are at times more heavily influenced by "folk" wildlife management strategies than the research of Fish and Wildlife biologists, creating a highly politicized performance of contrasting sensibilities regarding human–nature relations and the validity of Western science. Perhaps the most noteworthy of such social dramas involved the approximately thirty-year debate over the appropriateness of antlerless deer seasons to thin the burgeoning deer herd; a management technique favored by biologists, but often opposed by rural Vermonters.[19]

2

A Discourse of Interdependent Human–Nature Relations

> It's just part of life. That's something you learned young. Some things had to die so other things could survive.
>
> Man from East Middlebury

Throughout my fieldwork, during both interviews and participation in everyday activities, I was reminded of the extent to which many rural Vermonters lead lives that are characterized by a consistent mental and physical engagement with their physical environment.[1] One of the most unforgettable examples occurred one night during the opening weekend of the 1997 deer season at a Rochester bar called the Silver Tooth.

The Silver Tooth was packed with men, women, and a few children eager to celebrate the opening of another deer season. There was a decent band playing standard barroom covers, and many of the men were still wearing their camouflage hunting clothes. Surveying the scene around me, I felt incredibly lucky that I had tagged along with some of the younger guys instead of staying at the deer camp. What a wonderful glimpse this was into the esteemed social position of deer hunting in a small Vermont town. Little did I know, however, that an event would soon occur that would leave me feeling as if I had stumbled into the mystic heart of Vermont deer hunting and witnessed something outsiders could never even imagine.

The band was in the middle of an impressive rendition of the Doors' classic "L.A. Woman." They were repeating the verse, "I woke up this morning and got myself a beer," when suddenly—*amazingly*—they began singing "I woke up this morning and *shot myself a deer*." And, as if to perfectly complete the scene for the dumbfounded anthropologist choking on his beer by the pool table, the crowd began singing along at full-throttle—smiling men and women pumping their fists in the air and singing "I woke up this morning and shot myself a deer." For several seconds time stood still for me as I tried to process something that seemed beyond the realm of imagination. Then I laughed to myself, realizing that of all the places I could be on Earth that night, this was about as good as it was going to get if my intention was to learn a thing or two about hunting in rural Vermont.

The hunters I have met in Vermont generally approach their physical surroundings in a rather practical manner, clearly designating things such as trees and animals as resources that can and should be utilized by human beings.[2] For example, an East Middlebury man who grew up farming (not the one quoted in the epigraph to this chapter) told me, "I was exposed very early to the fact that animals become food for man." A Pittsford woman involved in local politics said, "How can you go out and shoot Bambi? Simple, it's dinner." A young woman from Lincoln who had been involved with hunting all her life put it concisely: "Some animals kill other animals."

Yet the fact that animals are natural resources did not prevent hunters from expressing deep respect, sometimes even affection, for them. Furthermore, most hunters I met viewed hunting as an important aspect of a larger stewardship process in which they were caring for their physical environment. So while hunters thought humans—like all animals—were dependent on the physical environment for their ultimate survival, they also felt that humans, in turn, are necessary to maintain the health of the physical environment itself. I refer to this dynamic interplay between humans and their physical surroundings (articulated in both words and actions) as a *discourse of interdependent human–nature relations* This perspective bears a clear resemblance to the kinds of human–nature relationships we see again and again among indigenous hunter/gatherers all around the world. There is, however, one critical difference (among others, of course) between the hunters I know in Vermont and hunters in places like the Arctic Circle or Amazonia: it is extremely rare for Vermont hunters to discuss their relationships with animals, or "the environment," in spiritual terms. More specifically, supernatural figures or forces were never granted any role in the maintenance of either the health or size of animal populations.[3] For example, the idea of spiritual "gamekeepers" who regulate game populations is a fairly common one in the anthropological literature. Most Vermont hunters speak instead in terms of a folk wildlife management model (as mentioned in chapter 1), mixing wildlife management teachings with local experience *and* local truisms involving "Mother Nature." In fact, Mother Nature is probably the closest thing to a supernatural intermediary ever mentioned by the hunters I have encountered, though she is clearly a metaphor for natural processes, rather than an actual spiritual figure.[4]

Despite the fact that hunters in Addison County care deeply about the health of the natural world, we should remember that their thoughts and practices are decidedly anthropocentric—oriented toward the benefit of human beings. And I want to emphasize that, in opposition to the tendency among too many anthropologists to make overly broad generalizations about "the West," I do not claim that my Vermont findings regarding spirituality and hunting are necessarily applicable to all regions of the United States. For example, I would not be surprised if research in the southeastern United States yielded contrasting data on this point; particularly in light of the fact that the governor of Alabama, Bob Riley, declared one week during the summer of 2007 "Days of Prayer for Rain" and "asked citizens to pray individually and in their houses of worship" (Riley 2007). This is precisely the kind of supernatural agency that I do not hear about in Vermont.

Rural Practice

Among the hunters I studied in Vermont, ideas pertaining to the proper interactions of humans and animals are strongly influenced by a rural lifestyle. While this may seem an obvious point, it helps to explain the great disparity between attitudes toward the nonhuman world and wild animals in Addison County and the attitudes one ordinarily encounters in urban and suburban America.[5]

In the normal course of their daily lives, rural Vermont are exposed to mundane existential realities that most people in suburban and urban settings can live their entire lives without ever witnessing. During both interviews and casual conversations, the influence of rural life was often mentioned in direct connection with ideas about hunting, specifically with regard to a prevailing acceptance of the idea that the lives of animals (including humans) often depend on the deaths of other animals.

My intention here is not to romanticize life in rural Vermont. Because Vermont has become so strongly associated with wholesome rural life, I am wary of coming across as yet another person singing the praises of Vermonters' rugged individualism in the face of an advancing modern world. On the contrary, I am merely attempting to point out that life in rural Vermont (and rural locales in general) differs significantly from life in an American city or suburb. Understanding the ways that rural Vermonters engage their physical surroundings in the course of everyday life is critical to understanding how Vermont hunters relate to animals and, consequently, the meanings associated with hunting.

Of course, all of this is not to say that simply living in a rural area causes people to think and act in specific, predictable ways. There are far too many other cultural and historical contingencies at play to make such a simplistic claim. But life in a rural area undoubtedly provides people with opportunities to engage the world in ways that may lead them to think and act differently than people from more urban locales. Recent "red state/blue state" analyses and reams of sociological data leave little doubt that there are some important and predictable differences between people from rural and urban areas, such as the much greater likelihood of rural residents to consider hunting an ethical endeavor.[6]

Nature as a Natural Resource

Several key ideas about human–nature relations were expressed by most of the hunters with whom I spoke. These ideas are directly related to the everyday rural practice of generations of Vermonters and form the foundation of a cultural understanding of hunting in rural Vermont. The first of these reoccurring ideas was that "nature" is a *natural resource*. The exploitation of the physical environment world is thought to be a self-evident necessity of life, and therefore ethically unproblematic. Many people during the course of my research—especially those in their mid-forties and older—explained their real dependence on hunting for food when they were younger. For example, one night at a deer camp in Rochester the senior member of the camp told me, "I can still remember when getting a deer was very important to our family." In the 1940s and 1950s, hunting, for many families, was a year-round pursuit. And though most people I talked to no longer felt this kind of pressure to have a successful hunt, the stories still persist; along with the confidence that the forest around your house can provide for many of your basic subsistence needs.

A watershed event in my fieldwork occurred in the fall of 1997, when I was walking along the trail to Mt. Abe in Lincoln with a woman from Bristol whose family has lived in the area since the mid-eighteenth century. After eating our lunches above treeline on the top of the mountain, we began walking back down at a nice leisurely pace while discussing some local environmental issues. Eventually the conversation began to revolve around the differences between the "native" Vermonters and the more recent Vermonters who have moved to the state, in most cases to escape overcrowding somewhere to the south. The woman I was walking with (a college-educated, nonhunting, outdoor-loving artist) expressed frustration with some of the attitudes that she noticed in many of the more urban environmentalists—specifically, the sort of *hands-off* approach that often seems to be encouraged by these people. As one who grew up around farming and lumber mills and depends on the wood that comes off her own woodlot to heat her house in the winter, the condemnation of engaging the natural world and utilizing it as a resource seemed to baffle her. Finally, she looked over at me and said, in a voice that communicated both exasperation and confusion, "To me, a tree is heat!"

Of course, I was well aware of the dependency of many rural Vermonters on trees for heat before I went on that walk.[7] Nevertheless, I had not quite realized the extent to which some people apprehend the various components of the nonhuman world—trees, animals, rivers, and so forth—in terms of their value as natural resources. As I have written elsewhere, "It is important to keep in mind that, in the eyes of many Vermonters, they do not only see cute animals and pretty trees when they look out their windows. They also see meat and heat" (2000, 20). Actually, they often see both. And in no way does exploitation necessarily preclude affection.

More specifically, most of the hunters viewed the nonhuman world as an indifferent provider. Although people in Vermont do not live off the land to the extent that they once did, most people I encountered were quick to point out that we humans are still beholden to natural processes for our existence, regardless of how far removed from the natural world we appear to be. Furthermore, Mother Nature does not particularly care whether we survive or not. In this regard, wild animals and humans share a great deal in common in that they are involved in an overarching process that requires the death of certain organisms, both flora and fauna, to sustain the lives of others.[8]

Respectful Exploitation

Hunters almost universally expressed great admiration and affection for the animals they hunted. For example, one landscaper and avid bow-hunter remarked: "I love deer. I think they're a great animal. Maybe it's as much respect as it is love." Another man who lived only a few miles down the road said: "I love deer. To me there's an awe about 'em. I like watchin' 'em. I like everything there is about 'em." Statements like these are particularly noteworthy because they provide evidence that contemporary hunting among Euro-American Vermonters is not necessarily associated with hostility toward animals. On the contrary, hunting often seems to enhance appreciation of game animals, albeit in an anthropocentric way. So while it may be a fact of life for many rural Vermonters that animals must die for human survival, this is not generally associated with a lack of respect for game animals.

The respect Vermont hunters have for their prey seems to emanate from three basic sources. First, throughout Vermont's history people have largely lived in rural areas. They realize their dependence on the nonhuman world and they feel they play an important role in maintaining its health. Second, the hunters generally believe in the existential necessity of animal deaths as part of the natural process of life. A hunter and artisan from New Haven Mills, while commenting on the unpleasantness she feels about the actual deaths of deer, spoke directly to this point when she said, "I love deer. I think it's just nature and we're part of it."

The third key element of the respect felt for animals derives from the fact that successful hunters generally possess a considerable amount of general and specific knowledge of the animals they hunt. On a general level, hunters must understand the typical habits of their prey, such as where it lives at different times of the year, what it eats, its life cycle, and its temperament. This basic knowledge of what particular species do to survive in the wild usually results in a heightened sense of respect for the animal.

Hunters also gain a great deal of knowledge about animals through pre-hunting-season scouting. Generally, but not exclusively, associated with deer hunting, scouting is heading off to a probable hunting site before the season in an effort to learn about things like animal locations, movements, and food availability. In this way, hunters often possess intimate knowledge of specific individual animals long before the opening day of hunting season. A hunter may know where, when, and what a certain buck eats, where he spends his nights and where he spends his days. Essentially, the hunter gets a glimpse into the unique daily habits and "personality" of an animal, and this increases the chances of actually killing it once the season opens. One hunter, a contractor from Pittsford, summarized the process:

"The best way you could ever figure out how to hunt a deer is to be with 'em all the time."

Animal Deaths

Just as a natural-resource orientation toward animals does not diminish the hunters' respect for them, neither is the actual killing of animals taken lightly by the great majority of hunters I know in Vermont. As a result, most of them expressed mixed emotions about the killing involved with hunting. While they were happy with their success when they killed an animal, they were also saddened that a beautiful animal was dead.[9] As one man in East Middlebury put it, "I feel extremely elated and also feel a little bad." Another East Middlebury man said: "It's exciting right up to the point when you walk up to it and look at it and say, 'Well, that animal was alive a few minutes ago.' That part always bothers me." One woman's first words when I asked how she deals with a dead deer were "I always bawl."

In fact, some hunters felt bad enough that they quit hunting; others were considering quitting. One man in Salisbury, who told me that he killed deer for food all year round when he was younger, explained that he does not hunt anymore because he feels he has done enough killing. Another man, in his forties or early fifties, told me he might stop killing deer as well and become a "picture-hunter."

Several men and women told me that they had been severely affected by "messy" kills at one time or another; enough so that some considered never hunting again. A woman in Lincoln—only in her early thirties when we met in the late 1990s—who is a very successful hunter told me that as she gets older the killing is getting more difficult for her to handle: "I don't like the killing part. I really don't. I think each one I feel worse and worse—each deer I kill." To help place this statement in the context of our conversation, she was not saying that she feels killing a deer is morally wrong. She was referring to the bloodiness of the whole affair. As she gets older, she finds that she has less tolerance for the goriness of the situation.

Once again, as I have been reiterating throughout this book, the reason why the killing is considered acceptable in the first place is that it is thought to be a natural, though sometimes unpleasant, aspect of life. For example, I spoke to a woman from Bristol who hunts with her family and loves eating venison, but also refuses to actually shoot anything, because

she is afraid she will only injure the animal and cause it to suffer unneces-
sarily. "They're so beautiful and everything," she said. I then asked her,
given the fact that she is so concerned about the suffering of deer, how
she can condone the killing of them at all. "It's part of life," she replied. " I
don't know how people can live in cities. It's like growing vegetables."

During the course of my research in Vermont an important seman-
tic distinction came into focus: *guilt* versus *sorrow*. I do not have record
of a single hunter telling me that she or he felt guilty about killing wild
animals. They did, however, as this section demonstrates, admit to feeling
sorrow. The key difference, of course, is that guilt implies doing something wrong and sorrow does not. This distinction goes a long way toward
explaining how most of these people thought about their relationships
with the nonhuman world. It is a relationship that is inherently imbued
with a certain amount of sorrow because it includes the death of beautiful and respected creatures.

Hunters as Stewards of the Earth

These environmentalists are clueless—absolutely clueless! Their idea is
that if you leave everything alone the world will be perfect.

Man from Bridport

Up to this point, my ethnographic description of interdependent human–
nature relations has been unilateral, as I have focused on the ways that
hunters in Vermont take resources from the nonhuman world through
respectful exploitation. At this point, however, I would like to examine
the giving side of this interdependent relationship: stewardship. There is
a widely held belief among hunters in Vermont that humans must take
an active role in tending to the general health of the planet, and especially
to the health of animal populations. Indeed, it was the rare interview that
concluded without at least a passing reference to the practice of stewardship.[10] I concluded from my interviews that many hunters actually felt that
hunting and killing animals is truly a way of caring for the nonhuman
world.

Unbalanced Nature

The single most important idea to understand with regard to stewardship practices among hunters in rural Vermont is that most hunters do
not subscribe to the popular notion that, left to their own devices, the
flora and fauna around them are capable of reaching a state of harmoni-

ous equilibrium—the so-called balance of nature.[11] In the opinion of most Addison County hunters, the natural world is characterized by dramatic swings in which various animal populations are always liable to become imbalanced for one reason or another, causing problems not only for certain animals but for humans as well.[12] As a result, anything resembling the idealized "balance of nature" is considered to be the result of intentional human wildlife and forest management practices.

Far and away the most common example provided by hunters testifying to the need for human intervention in natural processes was rabies. Indeed, rabies was mentioned by nearly every hunter I interviewed. Generally, a rabies outbreak occurs among animals that can be trapped for fur, such as the fox or raccoon. The reason for the outbreak, according to the hunters, is an overpopulation of the infected animal caused by a human failure to keep the population in check. Often, as is the case with foxes, these animals are predators themselves, so they have few nonhuman predators to thin their numbers. The general feeling of hunters is that, left unchecked, these animals will overpopulate. An East Middlebury man reported: "The way I explain it to my daughter is man was put on the Earth—everything was put on the Earth—for a reason. If we don't control the population of animals Mother Nature's gonna. And when Mother Nature does it it's a lot more cruel than the way we do it." A man just down the street said: "If you don't take care of 'em then Mother Nature's gonna. That's a plain simple fact."

In a discussion of these issues with a Middlebury woman, her husband, and her father, the woman spoke definitively on this issue and explained that when people "are taken out of the food chain" animals will become unhealthy. Not surprisingly, she described rabies as "Mother Nature's way" of cleaning up the problems of animal populations.

Two more examples of the negative effects overpopulation that I often heard were only mentioned in reference to the white-tailed deer population. One was a generally unhealthy population due to a lack of available food and the degradation of the gene pool. The other was simply starvation as food supplies ran short over the long Vermont winter.

After conducting numerous interviews and becoming well acquainted with this stewardship ethic among hunters, I noticed what I considered to be a persistent logical inconsistency: hunters routinely told me that animal populations required human management to maintain any semblance of balance, but rarely made mention of the problems that have been created by the human eradication of large predators, such as wolves and mountain lions. Were not these hunters, in fact, conveniently neglecting to mention that we humans—especially in eastern states like Vermont—had annihilated a crucial mechanism of wild animal population control?[13]

What I finally realized, however, is that this apparent contradiction depends on the assumptions with which one approaches these issues. For example, the critique of the hunters' omissions of large carnivores from their conversations about the "natural state of nature" rests on the assumption that nature would, in fact, be balanced if large carnivores existed in sufficient numbers. If, on the other hand, one takes the approach that nature does not ordinarily reach such a state of functional equilibrium, then it could be reasonably argued that humans should be "stewards of nature" regardless of what other animals are present. Of course, from this perspective humans are thought to rightfully, and unapologetically, rule the animal kingdom.

Learning from Coyote

Sometimes overpopulation results in the domination of one or more species by another, such as in a lopsided predator/prey relationship. The concern over the Vermont coyote population is probably the best example of this scenario. For instance, it is common knowledge among Vermont hunters that the rabbit population has dropped precipitously as a result of an increase in the coyote population.[14] More irksome to most hunters (because most are more serious about hunting deer than rabbits) is the effect of coyotes on the deer population. I have heard many hunters say something like "One coyote is one less deer." Coyotes are also blamed for attacking deer in a way that causes excessive suffering, and for killing them for the sheer joy of the experience. These concerns are at the root of the prevalent belief in rural Vermont that deer are more justifiably killed by humans than by any other predator.[15] Moreover, most of the hunters I interviewed displayed a possessiveness of the deer they hunted, which often manifested itself in a competition of sorts with coyotes. While the white-tailed deer is highly respected among hunters in Vermont (sometimes to an extent verging on reverence), the coyote is an animal that many treat as vermin and shoot on sight in an effort to protect the cherished deer herd. Ultimately, both animals are liable to end up in the crosshairs of a hunter's rifle, but for dramatically different reasons.[16]

I have already provided some examples of the great respect hunters express for the white-tailed deer's ability to survive in what is generally considered to be an extremely challenging physical environment. In this lone respect, opinions on coyotes and deer are actually similar. Indeed, very few people who spend much time in the woods can deny that the ingenuity of the coyote is quite impressive. As one man in Rochester said of coyotes, "I don't like 'em, but I really admire 'em." This man also conceded that if he had a greater understanding of the coyote there would be a good chance he would harbor less ill will toward these animals. Precisely the same sentiment was touched on by a man from Pittsford: "I just don't like 'em. I don't like 'em. I don't know if it's more fear than anything else."

Much like wolves throughout world history, coyotes are viewed with great ambivalence in contemporary rural Vermont. And this ambivalence is associated with more than simply their threat to deer; there is a direct personal fear of coyotes as well. A man in East Middlebury displayed this mix of fear and respect: "Another thing about coy dogs—they kill for no reason. They don't kill 'cause they're hungry, they kill for no reason. . . . They'll kill just for the sport of killing. To me, they're just a big unkept dog is what they are. Just overly grown and just a wild dog. A lot smarter than a normal dog—I mean wild smart."[17] He continued with an especially revealing comment as he gazed out his kitchen window: "I think deer is the most beautiful thing that you could see out in that meadow. You don't have to worry about one attacking you. . . ."[18]

While I was conducting research in the spring of 2007 on the controversy over coyote-hunting tournaments in Addison County, a tournament organizer expressed both the respect and the fear that is often part of discussions of coyotes in rural Vermont. "I admire 'em," he said. And then he went on to explain, quite adamantly, that coyotes need to be hunted, not only to protect deer, but to protect people by preventing coyotes from completely overcoming their fear of humans. As he put it, "You gotta keep them nervous." Another enthusiastic supporter of coyote tournaments, a local United Life Church minister, offered his perspective on coyotes: "They're a part of the whole creation, but they have to be limited."

A comparison of common sentiments regarding deer and coyotes provides insight into the idealized vision of human–nature relations held by rural Vermont hunters. In general, the people I encountered in Vermont expressed great appreciation (in both a utilitarian and an aesthetic sense) and respect for animals like deer, rabbits, and grouse—all relatively innocuous animals that are considered good to eat—and a relative lack of affection for predators like coyotes and wolves, which were generally thought of as potentially dangerous, nonedible competitors that needed to be kept in check for the sake of both animals and humans. These two opposing groups of animals can usefully be categorized as the "manageable" and the "unmanageable." (We've seen that in the late nineteenth century deer were widely maligned for their propensity to damage crops, but deer, unlike the coyote, can be managed.) It's interesting to think about how these two categories relate to another, more widely used, dichotomy: wild and domestic. It appears that manageable and unmanageable are positions on a continuum of domesticity and wildness, positions defined by the extent to which the animal can be controlled (or managed) by human beings.

Because of their well known ability not only to survive human attempts to squelch them, but to actually increase their population in the face of intensive management by humans, coyotes might be a candidate for the wildest of the wild. Indeed, coyotes have been shown to as much as double their rate of reproduction when they are actively managed (Voigt and Berg 1987). Simply put, coyotes are the most uncontrollable and transgressive wild creature in the woods of Vermont—an anxiety-provoking potentiality of chaos. One need not venture out onto a thin limb to be struck by the symbolic importance of the encounter between domestic, human-obedient hunting dogs as they sink their teeth into the flesh of their wildly disobedient cousin—the "big unkept dog," "overly grown" and "wild smart"—the unvanquishable coyote. And unlike wolves, coyotes seem to flourish in fairly domesticated spaces. One even ended up in New York City's Central Park in 2006. So not only are coyotes unmanageably wild, they bring their wildness into the domestic sphere, aggravating people all the more.

Meanwhile, as important as symbolic analysis is to understanding the coyote in Vermont, we should not forget about the material reality of the coyote, or perhaps I should say the so-called coyote. For decades, locals have been discussing the strange characteristics of coyotes in New England, claiming that they don't adhere to the accepted norms of coyote size and behavior. They're about ten pounds heavier than western coyotes (Wilson, Jakubas, and Mullen 2004, 4), and they appear to be running in packs, which makes them seem like a greater threat to both deer and people.[19] Biologists wondered about the same things. And they recently solved the riddle of the animals that many locals call "coy dogs" or "coy wolves." These perplexing canines are actually coyote/wolf hybrids. More precisely, they are crosses between coyotes (*Canis latrans*) and eastern Canadian wolves (*Canis lycaon*) (Wilson, Jakubas, and Mullen 2004, 2).[20] This new information has led some biologists to give these animals the mock-Latin name *Canis soupus*, because their DNA is all mixed up like soup.

The new data on the genetic hybridity of these animals explains the larger size of the coyotes in New England in comparison to western coyotes. It is not unusual for hunters and trappers in Vermont to kill coyotes that weigh thirty-five to forty pounds, and there have been exceptional individuals that have weighed upwards of seventy pounds. I saw a pelt of this size-class in an East Middlebury man's basement, and I couldn't help but make a mental comparison to wolves. One somewhat famous wild canine was shot and killed on October 6, 2006, by a dairy farmer in Newport Center. It weighed a staggering ninety-two pounds. The federal government confiscated this animal to determine its origin, and about a year later the public was notified that it was, in fact, a wolf. Interestingly, based on a number of genetic and physical characteristics, researchers concluded that the huge male was "almost certainly bred in captivity" (Hirschfeld 2007, 1). Whether this wolf was out in the woods by accident, was part of some private citizen's personal wolf reintroduction project, or was actually a wild wolf is unknown.

Habitat Enhancement

Keeping populations in check through hunting and trapping is one of the two ways of practicing stewardship that I observed during my fieldwork; the other is habitat enhancement. Many hunters actively manage their land in an effort to create optimal habitat for deer growth. For example, some plant soybeans specifically to provide ample high-protein food for deer. A similar habitat-improvement strategy was practiced by many Native American tribes as well, usually by burning off mature vegetation to provide deer with fresh young browse (Sutton 2000; Krech 1999).

Anthropocentric Human–Nature relations

My informants' interactions with the natural world, though clearly respectful and involving active stewardship, are also unambiguously anthropocentric. To these hunters, the natural world is something that is engaged, utilized, and managed in an attempt to satisfy human needs. Once again, when I refer to interdependent human–nature relations I am not implying human–animal equality, balanced reciprocity, or spiritual harmony, as one might expect when a term like "interdependence" is used to describe human–animal relationships among indigenous peoples. None of my informants described animals as spiritualized beings, and none claimed that his or her personal spiritual life had anything to do with hunting success, or that supernatural figures or forces have anything to do with regulating animal populations.

Human–Nature Relations and Gender

The general perspectives on human–nature relations of the male and female hunters I met in Vermont were essentially identical. Although this consensus may come as a surprise to some readers, it is merely an affirmation of the main conclusions that sociologists have reached on the subject. One would expect hunters, male and female, to share a great deal in common regarding their views of human–nature relations, particularly if they are all from similar social environments (Kellert 1976; Heberlein 1987).

The great majority of hunters I met in Vermont, regardless of sex, shared the views that it was quite natural for humans to kill wild animals, that the nonhuman world is a resource, and that it is imperative to eat what one kills; most also felt great affection for the nonhuman world. Nor was there any noticeable difference among males and females in their interest in trophy hunting or hunting exotic animals; and both men and women commonly mounted the heads of animals on the walls of their homes. Both men and women described a certain "rush" of exhilaration when they killed animals, particularly large mammals. Often this feeling was difficult to explain, as one woman's attempt testifies: "It's just, I can't explain, I don't understand it." And, lastly, both women and men felt a certain amount of sorrow over killing animals, and some even were getting tired of killing.

Beyond their ethnographic value as comparative descriptions of male and female hunters in Vermont, these findings serve as an interesting contrast to the contention, found in both popular and academic circles, that women are less likely to be interested in hunting because of a "natural" feminine empathy for the nonhuman world. This position is most clearly articulated by certain theorists who group themselves under the umbrella of ecofeminism, an approach that emanates from the general contention that "there are important connections between how one treats women, people of color, and the underclass on one hand and how one treats the nonhuman natural environment on the other" (Warren 1997, xi).[21] In its more radical renditions, as Mary Zeiss Stange points out in *Woman the Hunter*, ecofeminism makes the intriguing claim that "the female consciousness has a close affinity with that of nonhuman animals, leading women to be able to communicate with and understand nonhuman animals in ways men cannot" (Stange 1997, 73; see Collard and Contrucci 1989). Moreover, women are said to possess "innate nonviolence," and an "aversion to weapons and hunting" (Stange 1997, 73).

Though they clearly appeal to many people, these characterizations of women are strongly contradicted by the ethnographic record, even in cases when women are not living under patriarchal rule (see Lepowsky 1993). Furthermore, in his outstanding review article "The Sexual Division of Foraging Labor: Biology, Taboo, and Gender Politics," Robert Brightman refutes the assertion that women are not psychologically suited for hunting by pointing out that hunting is quite common among women in hunter/gatherer societies: "Hunting small game is commonly included in women's foraging labor; unless prey *size* articulates in some fashion with the purported sex-linked differentials in aggression, it is impossible to derive the division of labor from innate dispositions" (1996, 695). Ironically, with respect to the stance of radical ecofeminism, women in hunter/gatherer societies are often responsible for dealing far more brutal deaths to game animals than their male counterparts, because they are very often prohibited from access to sophisticated weaponry. As a result, men often kill animals at a distance with arrows, spears, and bullets, while women are left to club or stab animals to death at arm's length (706). As we will see, the troubles female hunters have with hunting are not with animals or any innate female aversion to killing; they are with other people.

Ecological Occidentalism

It is important to recall here that anthropological comparison necessarily revolves around a fictive version of Euro-American culture, and thus this problem exists for all anthropologists regardless of their specific cultural backgrounds.

HENRIETTA L. MOORE, *Anthropological Theory Today* (1999)

On a variety of different fronts, my experiences among Vermont hunters have not agreed with the received wisdom offered up by either academia or popular culture. Nothing exemplifies this trend better than the disconnect I encountered between the human–nature relations I observed in Vermont and what the anthropological literature told me I saw. As I delved into the library stacks to help me understand the cultural practices of rural Vermont hunters and place them in cross-cultural context, I soon realized that there was actually very little space for Vermont hunters in the dominant anthropological perspectives on human–nature relations.[22] Among a surprising number of respected anthropologists, there exists a casual acceptance of an over-generalized Western view of nature, often referred to as "dualistic"—meaning that Western human–nature relations are dependent on a rigid nature/culture distinction. This approach, it is argued, stands in stark opposition to the human–nature relations of indigenous peoples (especially those who hunt), which often stress "continuity" with one's physical surroundings, with humans being just another set of inhabitants on a planet they share with a wide array of "nonhuman persons" (see, for example, Nadasdy 2003; Smith 1998; Descola 1996; Bird-David 1993; James 1993). While I certainly applaud any effort to unchain ourselves from the shackles of Western philosophical assumptions and to try to understand "others" on their own cultural terms, I also believe that the Western/Indigenous distinction commonly drawn in ecological anthropology is slipping into the realm of caricature. Indigenous people are too often portrayed as a harmonious part of the physical environment and Westerners are too often portrayed as antagonistic invaders of that environment. I call this approach ecological occidentalism.[23]

My data, however, is populated by people who seem to split the difference between the extreme poles of this binary scheme. They generally see themselves as "part of nature" and genuinely care about animals, yet they are also unapologetically utilitarian in their interactions with the other life forms around them. Not only does this Western/Indigenous binary basically deny that people are actually doing what they are presently doing in rural Vermont (acting toward animals in ways that are simultaneously consumptive and respectful), it also severely limits the possible complexity of the meanings associated with Euro-American human–animal relations in general.

At the most basic level, these stereotypical descriptions of "the West" are rooted in Primitivist philosophy, which—in contrast to Progressivism—bemoans the loss of an idealized natural purity that is assumed to occur with the development of civilization. In *The Philosophical Roots of Anthropology*, William Adams writes: "For the Primitivist, the Golden Age was far in the past, and often at the very beginning of the world. Everything since has been a tale of increasing corruption of the originally pure state of nature, including Natural Man" (1998, 75).

More specifically, the perception of Western/Indigenous dualism rests on the shoulders of the darling of Primitivist philosophy, the Noble Savage—an idealized character that dates back to classical antiquity and can be found in the writings of Homer, Aeschylus, Virgil, and many others (Adams 1998, 80): "The Noble Savage is a compendium of all that is best in Man, physically, mentally, and spiritually. Physically, he is hardy, brave, and capable of extraordinary feats of self-denial; mentally he has a heightened aesthetic sensitivity and a love of nature and of harmony; morally he is benevolent and generous. He is also, either by reason or by instinct, a conservationist and an ecologist, appreciating his place in the grand scheme of nature and taking no more from it than his immediate needs require" (ibid., 79).

The Noble Savage trope, then, lies at the foundation of a deeply problematic Western/Indigenous dualism that routinely pits the earth-friendly, spiritually-mediated human–nature relations of indigenous societies against the exploitive, rationalized human–nature relations of a homogeneous West. In this symbolic arrangement, the Ignoble Westerner becomes the necessary logical counterpart to the highly romanticized and culturally constructed Noble Savage. In other words, the Noble Savage trope is really a commentary on a *relationship between two different kinds of people.* You can't have the Noble Savage without the Ignoble Westerner. They are opposite sides of the same coin, with the Noble Savage trope saying as much about the West as it does about the Noble Savage. Unlike the Noble Savage, however, the Ignoble Westerner has escaped the scrutiny of anthropologists, an oversight that threatens to seriously obscure our understanding of human–nature relations among nonindigenous peoples and indigenous peoples alike.[24]

Although this is clearly not the place for an extended theoretical discussion of this topic, I would like to at least touch on one issue that I think merits a bit more attention. When I read books and articles that offer comparisons of Western and Indigenous human–nature relations, I am consistently struck by how wide the cultural divide between these two categories has become.[25] The idea that there might be commonalities between some admittedly extremely different people seems to be beyond consideration. It seems to me that this sort of either/or position can only be maintained if one discounts the common *practices* that may exist among people from different cultures and, instead, makes comparisons based solely on ideologies, conceptualizations, or philosophies. In the case of human–nature relations, the most important cultural comparison that is almost always made is between the spiritualized natural world (often referred to as an animistic perspective) of indigenous peoples and the de-spiritualized rational natural world of Westerners. This difference (which, I might add, is not an absolute) seems to be enough to eliminate any real opportunity for dialogue or understanding between members of these categories. But when we bring actual practices into play (as we should be doing in any case because ideologies and practices are always interrelated), we begin to see points of similarity, areas where comprehensible cross-cultural communication and understanding can occur.[26] For example, can we not say that some similar practical orientations to the physical environment might exist between, say, a white trapper and an Inuit trapper? Or, more broadly, might there be some commonalities in these two people's general perspectives on the relationship of humans to the natural world? Or, exclusively in the realm of practical skills, would we not all agree that these two people could chat (even if through an interpreter) about the ins and outs of trapping, even though they come from different cultural backgrounds that offer widely divergent explanations for the actions of nonhuman animals? Or, more to the point of my research, can we *definitively say* that an indigenous subsistence hunter from the subarctic or a horticulturist/hunter from the Amazon and a dairy farmer/recreational hunter in rural Vermont will not share some common perspectives on human–nature relations drawn from their common interests, concerns, and experiences with their respective home environments? And, in the realm of cultural politics, what should we make of the fact that rural white hunters and indigenous peoples are misunderstood by environmentalists in very similar ways because they often approach their physical surrounding in a similarly utilitarian manner?[27] If we agree that these types of commonalities may exist, then what do we do with the Western/Indigenous binary? Without suggesting that we abolish it, I do think we should at least recognize the disservice it does to both indigenous and nonindigenous peoples, reconsider its explanatory utility, and seriously ask ourselves what is at stake if we reduce its role in anthropological theorization.

3

Hunting in Vermont *Now*

In the next five chapters I will discuss a variety of different reasons why hunting is such a meaningful social practice in Vermont today. I start here with the broad overarching theme that has emerged from my ongoing research: the relationship of contemporary hunting to the everyday pressures of modern living. While the issues I talk about in this chapter (such as attachment to place and self-sufficiency) could very well be discussed outside of the framework of modernity, I believe this extra contextualization is important since it provides insight into the continued popularity of hunting in the twenty-first century.

Because of the interplay of science and technology, international travel, global media connections, and capitalism that characterizes contemporary life in industrialized countries, many social theorists take it for granted that life today is extremely anxiety-provoking. Traditional religious and cultural explanations for worldly events are being challenged by modern "scientific" expertise that, in turn, is constantly being challenged by competing expert explanations. (Can somebody just tell me once and for all if eggs are bad for my heart?) Meanwhile, just as people are swimming—or maybe drowning—in modern ambivalence they are being presented with an unprecedented proliferation of choices about what to buy and how to live.[1]

The title of Anna Quindlen's novel *One True Thing* always come to mind when I think about these things. What's real in our lives; what is true? What can we depend on when the chips are down? Judging by the bloated self-help sections in bookstores, it seems that lots of Americans are having trouble looking in the mirror and answering these questions in a way they really believe.

Although rural Vermonters might seem to be occupying a leafy island of tranquility in an ocean of modern turbulence, they are, nevertheless, logging onto the same Web pages, watching the same television shows, enmeshed in the same global economy, paying off the same credit cards, and wrestling with the same modern dilemmas as any other American. So how, then, do the meanings of hunting in contemporary Vermont relate to these overarching conditions of modernity? This question is important not only for the intellectual purpose of trying to understand contemporary hunting from the perspective of hunters themselves, but also because for many antihunters the most fundamental and damaging critique of hunting is that it is unnecessary for survival in the modern industrialized world. From this viewpoint, hunting is a brutal anachronism that has long outlived its utility and leads to one simple and powerful question: *Why?*

I hope to provide some answers to this and other *Why* questions, but I would also point out that these kinds of questions are not the focus of my work, since what I am trying to do is share my admittedly imperfect understanding of the meanings of hunting practices in Vermont today. Moreover, *Why* questions already assume that hunting is a morally suspect endeavor and, as such, they make a lot of sense to people who feel that hunting and modernity are strange bedfellows, but not much sense to those who don't agree with this basic perspective on the characteristics of modern life.

Historical Continuity and Community

Through hunting, present-day actions are firmly embedded in the past and are imbued with a sense of perpetuity. The main reasons for this are the role of ancestors in the rural hunting community and the long-term relationship to specific landscapes that are forged through generations of hunting. As the sociological data show, most hunters come from hunting families. The skills of hunting are generally taught within families, being passed down from one generation to the next (Heberlein 1987). As a Middlebury farmer told me, "My father was a hunter and that's where I learned it from."

Certainly children do not learn hunting in school. These days, in fact, it is increasingly criticized in school. Elders, then, are the most respected gatekeepers of valuable hunting wisdom, the real points of connection with the "old days," and when it comes to hunters in Vermont the old days are thought of in glowing terms. Very often, then, old men are highly revered by younger men—even teenaged boys—for the wealth of knowledge that a lifetime of experience has given them. In this way, younger generations are constantly looking back to the past for guidance.

And, without a doubt, the middle-aged hunters I came to know generally thought it was a very positive experience for their children to have the opportunity to hear the stories of the older generations. Both mothers and fathers told me that hunting was a good, solid activity that would help keep kids out of trouble.

Not only are hunters usually home-grown, they generally do most of their hunting close to home. It is not unusual in Vermont for three generations of hunters to be hunting the same land on the same day. Old teaching young, young surpassing the old and, eventually, old men, barely able to walk, being helped out to familiar places where they can sit in the woods with a gun across their knees, grateful to have lived for one more hunting season. This may sound sappy, but I have been told several stories of elderly hunters being helped out to age-old hunting spots.

This relationship with land is highly specific at times, as with the young man I mentioned in the introduction who had little interest in hunting even a couple of miles away from his traditional family hunting grounds. Individual hunters often become emotionally attached to certain places, and hunting season just would not be the same if they were unable to return to their favorite spot. One woman, for example, told me that every year she and her husband hunt on the same land during the opening weekend of deer season, regardless of whether the chances of getting a deer in that place are favorable or not. At this point in their lives, this annual trip has become meaningful in itself.

Through the historical linkage of people with each other, and people with a place, hunting can become a critical element in the creation of a local community—seemingly another one of those human essentials for which so many contemporary Americans yearn. The game dinners, the opening-day breakfasts, the local papers with lists of successful hunters—these are all part and parcel of an autumn ritual season, if you will, that places people on common ground regardless of financial wealth or political clout and pays homage to a shared bundle of local rural values. In January 2007, a woman who had been one of my first informants back in 1996 poignantly explained the importance of hunting to a sense of local community in the context of a larger conversation about her college-aged daughter's emerging passion for hunting:

> [Daughter's name] was on the breaking edge of, you know, "Am I hunting cause it's cool and I like the boys and stuff or do I really enjoy hunting?" And I think that moose hunt really cinched it for her because it was a situation where it was crummy, crummy hunting and it happened just like the best hunting experience happens for you—when you least expect it, in a place

we hadn't scouted, a place none of us had been. We were at loose ends, like "Where do we go? We can't find a moose." . . . Every place we went with her thinking it was gonna be easy—no moose. And we were like—I mean [husband's name] was like, "I don't even know where to go now." And it's almost dark Sunday, we stumble on this beautiful bull, she shoots it, and darkness comes and then—you know, calling the [friends' last name] and old Walter comes . . . and the camaraderie of making pulling sticks and lashing eight men on a pull-line to get this moose out and, you know, just really recognizing the hunting community as it is. It makes me want to cry thinking about it [voice cracking, eyes filling with tears]. In the dark, standing around pick-ups in the dark, when it's all over—I mean you can't create that! There's no other way to create it.

On a much broader level, hunting often connects people to the vast overarching history of humanity itself. Hunting, to a degree that sometimes surprised me, is characterized as the most fundamental human activity. One woman described hunting as "primeval" in an attempt to explain its appeal. Thought of in this way, a hunting trip can serve to reconnect people with some ancient basic human essence. A man who was eighty-seven years old at the time of our interview expressed this man-the-hunter sentiment: "Well, mankind started by having to produce meat. If he didn't he went hungry. And it's born and bred into human nature to do that." Once again, hunters are looking back in time for the truest examples of what anthropologist Clifford Geertz called the "really real." The stark and sometimes mysterious confrontation with life and death is granted far more existential authority than the highly mediated lives that we generally experience today.

For many people, hunting connects them to phenomena that are beyond doubt and worthy of absolute trust. I think, for example, of all the people who described hunting in terms of a return to a better time or place. As one man in East Middlebury suggested, the "difference" with hunting is the "reverting back" that it requires. Hunting, this man said, is a matter of "feeding the soul." A man from Ferrisburgh echoed a common theme of hunters' testimonies when he succinctly referred to hunting as a "base" from which to carry out one's life, since it reaffirms basic existential realities that he felt our society is forgetting as it becomes more urban.[2] At a time when many people are concerned about being alienated from the most fundamental processes of existence, hunting permits (at least in the opinions of many hunters) full participation in these processes in an unfiltered way.

Depending on the individual hunter there are varying levels of self-consciousness, or reflexivity, involved with the matter of authenticity. There are some people who have "always" hunted and don't spend much time navel-gazing, and others who are more analytical, examining their lives, as if from afar, and wondering how exactly they fit into larger social and natural processes. The ultimate example of the latter scenario may have been the fellow who half-jokingly referred to himself as "a postmodern hill-hippy." At the time he was not a hunter, but he was seriously considering the idea of hunting in the near future. This man was in his late thirties, held a master's degree from the University of Vermont, and was working in rural community planning. Although he grew up in an urban area outside of Vermont, he was living in a cabin with no electricity and had no desire to experience urban living ever again.

I interviewed him one evening at the restaurant where I worked after I finished my shift. As the college kids filed in to drink at the bar, we went to a quiet table and talked. One topic in particular seemed to concern him more than any other: the contradictions he saw in his own eating practices. He is an ex-vegetarian who, at the time of the interview, had recently decided that it is acceptable, both morally and environmentally, to eat meat. But he was troubled by his own ambivalence about killing animals. Basically, it seemed dishonest to him. "If I didn't know if I could kill meat, if I couldn't deal with their killing, then I shouldn't eat meat . . . When I go hunting, I'm gonna dig the hunt, I'm gonna dig being in the woods, I'm gonna dig the whole predator thing, but I'm not gonna dig the act of killing. That just doesn't do anything for me. . . . I have a worldview, and in my worldview you understand your whole food system and you try to eat good food and, to be honest, I haven't decided if I'll be able to hunt. I just don't know. You know, if I find a nice buck and I know where that nice buck is—I'm gonna go out there and kill that nice buck? I don't know."

Regardless of the level of reflexivity involved, the hunters I met were in nearly universal agreement about the authenticity of the natural world. This basic core belief not only is critical in substantiating hunting as an ethical practice, but also helps explains why hunting is such a welcome existential alternative to the vicissitudes of modern life, and, ultimately, why hunting is so deeply meaningful and fundamental to the personal philosophies of so many people.

Self-Sufficiency

Not surprisingly, self-sufficiency is a common topic of discussion among hunters. Indeed, a New Hampshire hunter I talked to several years ago told me that the most satisfying part of hunting for him was the moment when he placed the butchered meat in his freezer. As one might expect, certain commentators have taken a somewhat skeptical approach to this issue, pointing out that this sense of self-sufficiency may be more accurately described as an *illusion* of self-sufficiency (Dizard 2003; Uzendoski 1978). Considering just how difficult it is to actually be self-sufficient, this critique is certainly legitimate. Nevertheless, regardless of whether the self-sufficiency is illusory or legitimate, the simple fact that people are yearning for it is significant from a social-scientific perspective. The hunters I spoke with in Vermont both desired self-sufficiency and provided some useful insights into the origins of this desire.

A common thread among the hunters I know in Vermont is a high premium placed on eating what one kills. Indeed, to most hunters the consumption of animals, in terms of keeping hunting within the bounds of ethical behavior, is probably the most important factor in the entire hunting process. The thought of eating clean, wild game that they had the most intimate of roles in obtaining was very satisfying to nearly all of the hunters I met. It is important to keep in mind that many of these people not only hunt, but also farm, keep large vegetable gardens, and forage for wild edible plants as ways of obtaining their food.[3] In each case, they expressed deep satisfaction in their ability to feed themselves without being overly reliant on store-bought food. Borrowing a phrase from anthropologist Uni Wikan (1990), self-sufficiency is a "compelling concern" among many rural Vermonters.

A woman I spoke to one afternoon at her house in New Haven Mills put words to what I presume are fairly common feelings among hunters.[4] The mother of two young daughters, she had been hunting since she was fifteen and placed a great deal of importance on self-sufficiency. One reason for this was that she felt it was important not to be a "dependent woman." Another reason was a dissatisfaction with what social scientists refer to as de-skilling. As she put it, "We depend on so many other things and I don't want that." It was very important for her to know that if all else fails (she mentioned her marriage and the economy as potential examples) she can still manage to feed her family. She works this philosophy into her life by producing a number of basic foods at home, such as potato chips, bread, and bagels, and by making an active attempt to increase her knowledge of many practical skills, such as automotive repair.

As for her daughters, she told me that she hopes they follow in her footsteps and grow up to become independent women. She said that she and her husband have agreed to teach them skills, such as carpentry, that will help them take care of themselves later in life. Lastly, she was determined not to be overprotective of her daughters in an effort to teach them personal responsibility. She felt that today's children—unlike those in generations past—are sheltered from participating in dangerous activities and therefore never learn to take their actions seriously.

This reoccurring theme of self-sufficiency clearly relates to, and is informed by, several important critical observations of modernity. Many theorists have commented on the capability of modern life to cultivate a sense of existential malaise. Much, but certainly not all, of this malaise is thought to be the result of two main influences: the capitalist mode of production and the increasing importance of expert advice in modern society. As Marx suggested, people are increasingly disconnected from the fruits of their labor and are therefore lacking a sense of accomplishment and self-mastery. In addition, the de-skilling that has occurred as a result of living expert-driven lives of "mediated action" has left people feeling dependent, confused, and vulnerable.

The suggestion here, which clearly resonates in my own research, is that people feel a disconcerting lack of control over, or a lack of involvement in, some of the most important aspects of their daily lives, not the least of which is material subsistence and the products of their labor. By becoming more self-reliant through hunting, gardening, and gathering, people are actually counteracting the effects of de-skilling and developing spheres of their lives that are intimately related to the fulfillment of their most basic needs. And while it is certainly the rare hunter in Vermont, judging from my experience, who is able to approach true self-sufficiency in terms of the production of food, this does not nullify the real contribution that a deer, some fish, a turkey, or a season's worth of grouse can make to the family food supply. Thus, unlike pursuits such as tennis, golf, sailing, or watching television, hunting is a *productive leisure activity that seems a little bit like work*. The most important point to keep in mind here is that, regardless of the extent of self-sufficiency one achieves, real production does occur each time an animal is killed and eaten. This much, at least, is not an illusion, and it carries great importance for hunters, as it gives them a small taste of freedom from what Max Weber called the "iron cage" of the modern world.

Herbert Applebaum's ethnography *Royal Blue: The Culture of Construction Workers* provides some helpful commentary on this issue. What Applebaum tells us about the working lives of construction workers stands in stark opposition to the conditions that many social theorists find so troubling about contemporary Western society. As workers in a handicraft industry, construction workers often supply their own means of production, work in relative autonomy, and exert a great deal of control over the workplace. In addition, the success of their work ultimately depends on the completion of entire projects, rather than random segments of a greater whole that they will never see. This set of characteristics is a far cry from the kind of work that most people do—whether in the white- or blue-collar sector—and makes the industry unique.

Significantly, construction workers score higher on job satisfaction surveys than any other occupational group. Applebaum reports: "Most construction workers believe they work hard, contribute to society, and earn an honest living. They feel they produce something real and tangible. They can see the physical evidence of what they accomplish, and the physical labor that tells on their bodies gives testimony to the integrity of their work" (1981, 22). I would submit that hunting provides similar satisfactions because it offers people a sense of productive freedom, personal satisfaction, and self-reliance.

An interesting corollary of the theme of self-sufficiency that I encountered in Vermont was the theme of frontier survival.[5] This current ran through a number of the interviews or was simply performed (obvious examples of cultural performance being people who hunt with blackpowder rifles reminiscent of the American Revolution and archers who are turning to "primitive bows" as their weapons of choice) by people who never spoke of it. One man used the term "homestead skills" to describe the outdoor skills he and his wife have learned over the years so they would be able to get by if the occasion arose. He and his wife actually celebrated Thanksgiving outdoors while I was in Vermont, cooking food that they had personally hunted or raised in their garden. A more dire version of the frontier survival theme related to the possibility of having to survive during a war or in the aftermath of a war. As one woman put it, "If Jim and I were the last two people on the planet, could we survive?"

With certain people, perhaps not surprisingly, there was a fair amount of frontier nostalgia associated with their hunting. A good example came from a farmer in Middlebury: "I was born a hundred years too late. . . . I'd rather live up on the mountain somewhere by myself than I would be in town. The city don't mean nothin' to me. I like bein' alone. . . . I would've fit in a hundred years ago very easily." A middle-aged woman from Vergennes echoed these sentiments when she told me that she would rather be living in the late nineteenth or early twentieth century because it was a simpler time that would have suited her love for the outdoors and her nonmaterialistic values (she told me she has no need for "trinkets"). One man, when asked about his fascination with bow-hunting, simply said, "To be able to do something like somebody did a hundred years ago."

Clearly, the idea of a highly skilled, independent survivalist who requires only the resources provided by the natural world stands in marked contrast to the alienated, de-skilled modern worker-bee so often characterized by social theorists. And, of course, this theme of frontier nostalgia fits very comfortably into the broader theme of authenticity.

Although all this talk of living off the land may strike many reasonable people as a frivolous and anachronistic fantasy, one must keep in mind that in rural Vermont there is somewhat more admiration accorded to people who are adept at skills that would increase one's chances of being self-sufficient—such as carpentry, logging, mechanics, agriculture, hunting, and fishing—than one would find in the suburbs. Indeed, especially for males, being unskilled in these areas can be judged as a sign of immaturity, even childlike helplessness. A particularly extreme example of this sentiment was voiced by a man who told me that if you can't build your own house then you don't deserve to live in one.

A more extended story of an incident that occurred at a deer camp serves as an excellent example of the high value of self-sufficiency and personal responsibility among many people in rural Vermont. As we were sitting around one evening at a camp in Ripton soon after I arrived, it suddenly dawned on one of the guys that he might know me from somewhere. Apparently I jogged his memory when I mentioned that I grew up in New Jersey. My status as a New Jerseyan, incidentally, was something that always concerned me during the early stages of my research, since in the minds of many Vermonters New Jersey is nothing but a big flatlander breeding-ground. I was always somewhat insecure, particularly at deer camps, of what people might be thinking when they looked out their window and saw the Jersey plate on my trusty Toyota Tercel wagon.

This particular guy, who had been known to take on a rather surly persona if need be, suddenly locked his eyes on mine and blurted out, "So you're the fucker from New Jersey who wouldn't give me a ride!" The year before, soon after I had begun my research, I had gone deer hunting alone in the mountains not too far from their camp. On that particular day, he had followed the same National Forest access road as me and was off in the mountains hunting very early in the morning—probably before daylight. Unfortunately, he left his truck lights on, and when he got back to it, miles from the nearest house, his battery was dead. As he was walking back down this road toward town, I was coming in the opposite direction, just starting my day of hunting. He flagged me down and asked if I had jumper cables. I did not. He kept walking. I thought I had asked him if he wanted a ride, but maybe I just remember it that way to make myself feel better.

In any event, now he was essentially indicting me of being a jerk from New Jersey who refused to help a Vermonter in need. Feeling startled, a little defensive, and somewhat confused by his tone, I jokingly said, "I'm not the one who left my lights on when I went hunting." Much to my surprise this comment, which was only intended to parry his initial charge, prompted an enthusiastic burst of laughter from the other guys that seemed to be saying, "He's got you there! Can't argue with that!" Not only was I surprised by the laughter, *I had won*. But now I was scared that this bear of a guy would be even angrier with me than he was in the first place. Then I was surprised again, and I learned a little something in the process. The razzing was over and this man seemed to accept the fact, somewhat sheepishly, that he was the one who had done something truly blameworthy that morning, not me. I already knew that self-sufficiency was highly valued among rural Vermonters, but this incident brought that point home like no other. Without intending to, I had called his self-reliance into question and there was nothing he could do to defend himself.

Commodity Culture

An insatiable, or at least unhealthy, appetite for commodities is often considered to be a hallmark of contemporary American life. People define themselves and express themselves through the things they buy. As John Fiske writes in *Reading the Popular*, "All commodities are consumed as much for their meanings, identities, and pleasures as they are for their material function" (1989, 4). People are continually trying to fulfill ever-changing industry-generated needs with commodities that actually *do* very little (Lash 1990, 40). As a result, our "needs" become less material, increasingly symbolic, more fleeting, and harder to satisfy.

The case of hunters in rural Vermont offers a contrast to this world of endless commodification. This is not to say that hunters in Vermont are somehow immune to the tantalizing allure of commodities. Like anybody else, hunters would walk into the store where I worked and stare in that "if only I had *that*" way at something way out of their price ranges.

And some hunters certainly do purchase extremely expensive equipment if they have the financial means to do so. The difference is that the equipment they buy—whether a pair of mittens or a new rifle—is still expected to do a certain job, and it is ultimately judged according to its performance. Unlike so many drivers of fancy European cars, no hunter I know would use an inferior gun just because it was expensive and prestigious. There's no prestige accorded to the owner of a rare, expensive gun if that person is not a good hunter. Once again, this is not to say that hunters are not aware of the latest technological advances, whether in ammunition, firearms, GPS units, binoculars, or scent-free clothing.

There is also a strong traditionalist pull among many rural Vermonters, which is to say that while they were still building personal meaning through commodities, the kinds of meanings were not those that anxiety-peddling marketing firms bank on. For example, I was surprised at the number of hunters who, in choosing clothing, actually rejected the advice of outdoor clothing experts in favor of outfits that could have been worn fifty years ago. Most noticeable was the preference for cotton long underwear, even though the "cotton kills" mantra has been chanted in outdoor circles for decades. Certain patrons of the store actually complained that we did not carry any cotton "waffle" long underwear—a product that has long been considered obsolete. Once again, even in clothing, many people clung tenaciously to the ways of their grandparents. To be fair, however, I must also say that the hunters I saw in action were expert dressers, skillfully manipulating layers of cotton and wool (a fabric of worship among hunters for its warmth, water repellence, and silence) to fend off brutal winter weather and remain dry and warm. After hunting with a number of people it was easy to understand why they saw no need to wear more contemporary clothing: they were not cold.

But, as with everybody else who chooses not to dress in the dark, there's more to the way Vermont hunters dress than practical necessity. The clothing they wear also expresses who they are, regardless of whether or not it keeps them warm. Plaid wool jackets and green wool pants offer great protection from cold, snow, and wind, but they also clearly communicate a strong message of rural traditionalism and all it entails. The kinds of clothing hunters wear is just one of many ways that hunting in Vermont is ritualistic. By this I simply mean that hunting is an expressive practice that occurs year after year, according to a familiar format.[6] As Stuart Marks writes in *Southern Hunting in Black and White*, "As with any ritual performance, to hunt is to demonstrate the potency of the past in an attempt to structure the future" (1991, 6). This importance of clothing in broadcasting a message of traditionalism came through loud and clear to me one day at a deer camp in Ripton. A couple of hunters who weren't members of the camp stopped by to chat for a minute as they passed by on their way out to hunt. Looking at them, I thought they couldn't have seemed more different than the guys at my camp. They were dressed in the Robo-Hunter outfits you can find in one of those dazzling Cabela's stores—outfits that make you look like a cross between an NFL linebacker and a Special Forces commando. I couldn't imagine the guys at camp slipping into skin-tight black long underwear made of some shiny space-age fabric, no matter how astounding its wicking properties were said to be. It just wasn't who they were. It was too sleek, too slick, too new. The guys at camp look like regular guys from town in wool clothing, not secret operatives on a covert mission.

Temporary identities, then, are not what Vermont hunters are primarily shopping for when they head to the store or open a retail catalog. Rather, they are looking for products that work. Much of this, I argue, relates to rural practice, because rural practice emphasizes "doing," particularly with regard to physical tasks. As I noted earlier, in rural Vermont there is still a real need for people who know how to cut down trees, not just people whose jobs take place behind wooden desks.[7] As a result, to most hunters the make of one's gun pales in significance to one's ability to hunt. Unlike the tradition of foxhunting in England, the great majority of hunting in Vermont (or the United States, for that matter) is not intended to be a display of wealth, and those few people who attempt to transform hunting in Vermont into a performance of class distinction are often mocked by other hunters.

Hunting as Leisure

Leisure is what people in capitalist societies do when the events of their daily lives are not constrained, in time and space, by the requirements of work (Fiske 1989, 83; Turner 1982, 36), and the contributions of leisure theorists can be valuable to the researcher of modern cultural productions. In the case of hunting in rural Vermont, a leisure studies approach has proven to be very useful because it highlights the connections between the desire for authentic experience and the alienating effects of capitalist society. For example, in his influential work *The Tourist: A New Theory for the Leisure Class*, Dean MacCannell argues that tourism is fueled by a desire for "authentic experiences" (1976, 101) and that sightseeing is "a kind of collective striving for a transcendence of the modern totality, a way of attempting to overcome the discontinuity of modernity, of incorporating its fragments into unified experience" (1976, 13).

On another level, leisure can also be thought of as activity that is carried out *against* the pressures of social control (Fiske 1989, 80). The anthropologist Victor Turner elaborates on the potential rebellion inherent in leisure activity: "Leisure is potentially capable of releasing creative powers, individual or communal, either to criticize or buttress the dominant social structural values" (1982, 37). Leisure, then, demarcates a special period of time when people are free to think the unthinkable; to play with new visions of the social world and the self. According to the cultural theorist John Fiske, "Leisure is essentially a time for self-generated semiosis, a time to produce meanings of self and for the self that the world of work denies" (1989, 82). Turner expresses essentially the same sentiments, though in a slightly different form, with a simple model that depicts leisure as providing people "freedom from" and "freedom to." Leisure provides freedom from work and other social obligations and the "freedom to enter, even to generate new symbolic worlds of entertainment, sports, games, diversions of all kinds." Furthermore, leisure provides "freedom to transcend social structural limitations, freedom to play . . . with ideas, with fantasies" (1982, 37). Although we often think of leisure activities as a frivolous diversion from "real life," it is really pretty serious business.

Hunting in rural Vermont fits comfortably under the heading of leisure as defined in this way. During my field research, hunting was continually juxtaposed to time spent at work. One carpenter in Hancock went so far as to refer to hunting as "anti-work." As he and many other men and women would tell me, hunting offers a welcome respite from the stresses of everyday life. Hunting, a construction foreman in East Middlebury told me, "is one of the very few things that totally relaxes me." Aside from work, "stress" was often talked about in terms of phones, commitments, kids, and bills. Phrases such as "peace of mind" were commonly used to describe the soothing effects of hunting. As one farmer put it, "Just to get away from all the pressure. As soon as I get to where I'm going I don't ever think about this place." Another man, a landscaper from East Middlebury, said: "To me, when I go out hunting it's a place I can go out and think. I can set there in the woods and it's very quiet, it's just the animals and I can do anything I want. I can do two, three days worth of work in my head when I'm in the woods and know what I'm gonna do the next few days.

It's a relaxation thing for me. That's what I really enjoy. There's a saying, 'getting back to nature.' That's what it does for me. I'm out there in nature. I can set in my tree stand and watch the squirrels and the chipmunks and once in a while you can see a fox or a coon or something. They don't even know you're there—they're doing their everyday little chores that they do."

Discussions of hunting as leisure, however, do not begin and end with comments about the relief of stress. Hunters commonly described a clear binary opposition between the everyday world and the world of hunting—whether their hunting experience consisted of a day in the woods or two weeks at deer camp. The theoretical importance of these descriptions is that they provide a solid foundation of data to support the argument that leisure helps people construct important alternatives to their workaday lives.

The first time I remember hearing a discussion of hunting that was framed by a series of binary oppositions was during a conversation with a woman from Lincoln. She described the appeal of hunting as largely stemming from a desire to get back to what she referred to as "hunter-gatherer" ways in an effort to relieve stress: "Getting back to nature. Getting in tune with it. Being a part of it." For her, the basic distinction came down to *indoors* versus *outdoors*. Work fell under indoors; hunting fell under outdoors. Additional elements of the indoor setting were people and noise. Hunting, on the other hand, involved a solitary and quiet outdoor setting. Moreover, it provided her with the opportunity to experience something natural, undoubtedly real, and more in tune with her personal emotional needs than what was available for her at work. In sum, one could diagram her ideas along the following lines:

nature : society
hunting : work
outdoors : indoors
solitude : crowded
quiet : noisy
peaceful : chaotic

This woman's description corresponded with many other hunters' sentiments about the appeal of hunting, though few elaborated on the various conceptual relationships to the extent that she did. She was not alone, however. For example, one evening at a deer camp in Ripton, a man surprised me when he compared life "on the flats" and life "on the mountain." In this case, since most people in Addison County live in the fertile Champlain Valley and all of the hunting camps are located in the heavily forested Green Mountains, "the flats" becomes a metaphor for everyday life. Two men at a different camp a little to the north also drew a similar distinction, this time between "the lowlands" and "the mountains." A man at a third camp got right to the heart of the issue of authenticity when he spoke of a distinction between "the real world" and "the created world." Not surprisingly, for him the "real world" of the outdoors was the place where one experienced life's most meaningful events. I thought it was both humorous and revealing that at this particular camp the men commonly used the term "re-entry" to refer to going home from camp.

Another aspect of hunting in Vermont—the deer-hunting season in particular—that fits nicely into the leisure approach is its festive atmosphere. This is because it reinforces the idea that hunting is something that occurs outside the expectations of everyday life.[8] A woman in Lincoln, for example, fondly remembered what she called the "holiday" atmosphere created by special meals and family gatherings during deer season when she was a girl. She was also emphatic in making the point that hunting season entails more than simply the days that one actually goes out and hunts. To her, and many others, it also includes all the time scouting for deer before the season and all the time enjoying the fresh venison after the season.

A man I met at a camp in Bristol Notch described the excitement he feels during the final days leading up to deer season as "like a kid waiting for Christmas." Another man—probably in his early seventies—nostalgically referred to deer season in the 1950s as "a national holiday" because everything in town was suddenly organized around hunting, including the widespread expectation that most male children would not be showing up for school. In a simple calendrical sense, hunting cannot help but be associated with holidays since the apex of hunting season—the rifle deer hunting season—begins two weekends before Thanksgiving and concludes the Sunday after Thanksgiving, thus commencing the holiday season that ends on New Year's Day.

If holidays are times when people gather together, with great anticipation, in order to celebrate a particular event or idea, then descriptions of deer season in Vermont as holidays may be more technically accurate than one might expect. After living in Vermont for several hunting seasons, the idea of referring to it in terms of a holiday seems perfectly appropriate. In terms of excitement, deer season actually surpasses Christmas for many people. It is almost as if half the men in town, and a smattering of women, are about to play in the World Series or the Super Bowl.

I had a ringside seat for this yearly drama as an employee of the largest sporting goods store in Middlebury. Normally only open until six or seven o'clock, we would stay open until midnight the night before the opening day of firearm deer season. And it was well worth it, both ethnographically and financially. "Madhouse" would be a reasonable term for what would occur on deer season eve as we raced around the store selling high-powered rifles and ammunition, pac-boots, rubber boots, wool clothing, fleece clothing, Gore-Tex-lined clothing, hats, maps, compasses, binoculars, tree stands, knives, backpacks, socks, camp stoves, jerky, camouflage clothing, hunting licenses, global positioning systems, walkie-talkies, hand- and foot-warmers, and a multitude of products meant to attract bucks in the rut. The excitement in the store might be described as a mixture of the pregame excitement athletes often talk about and the excitement of someone about to embark on an adventurous excursion. Finally, after nearly a year's wait, people were legally permitted to head out in search of the legendary white-tailed deer.

Why Not Yoga?

When the manuscript of this book was undergoing review, one of the press's readers expressed some reservations with the logic of this chapter: first, that I hadn't explained why hunting (as opposed to yoga, for instance) would be specifically suited to serve the purposes I've proposed, and second that hunting participation in the United Sates is declining. If hunting is so effective at buffering people from the storm of modernity why, the reader asked, is hunting on the decline? In short, the reader believed that I had presented a functionalist argument that lacked attention to local culture.[9]

I do not claim that the hunters I met are primarily hunting in order to ward off the pressures of modern life. Some did tell me that they find hunting important for that very reason, but many hunted just because they always had. I would suggest that an unintended consequence of this sort of unselfconscious hunting is a certain kind of existential security that does, whether the hunters know it or not, give them something to "believe in" (i.e., "nature") in these tumultuous times. So for some hunters the benefits that I discuss were unintended consequences; for others, they were a stated goal.

So why hunt, as opposed to participating in some other recreational activity, such as yoga? Rural traditions such as hunting have a long history in Vermont and are deeply meaningful to many people, particularly in comparison to yoga and other relatively recent imports. I admit, however, that yoga practitioners and hunters are dealing with some of the same overarching existential dilemmas when they engage in their respective activities. Where they often differ is with regard to their cultural sensibilities. Clearly, hunting is a more classically rural activity in the United States than yoga. And it is clear that there are some important distinctions between rural and urban value systems in the United States, especially in terms of engaging the natural world. The fewer people who grow up engaged in traditional rural practices (agriculture, logging, and so forth), the fewer hunters there will likely be. The reason, then, for the decline in hunting is not that it hasn't provided an effective means of dealing with modernity. Rather, it's because rural sensibilities are on the decline, and therefore there are fewer and fewer people who might consider going hunting in the first place. Ultimately, then, my argument is firmly rooted in, and dependent on, historically constituted cultural meanings as opposed to functionalist theory. What I have attempted to do in this chapter is to discuss some of the benefits of hunting in the late twentieth and early twenty-first centuries without assuming that these benefits necessarily cause people to hunt in the first place—the classic functionalist error of confusing the benefits of human actions with the ultimate causes of those actions. My conclusion, the only one my data would allow, is that sometimes benefits are part of the cause, but they are never the only cause. An accounting of rational benefits does not explain why one particular practice is chosen from wide selection of options that may well be functionally equivalent (whether we're discussing the quest for personal authenticity in post-industrial society or decisions by hunter/gatherers about how to fill their stomachs); only culture and history can answer these specific questions.

4

Ethics, Emotions, and Satisfactions
of the Hunt

In the previous two chapters I provided ethnographic data to support some general claims about hunting in Vermont: that a discourse of interdependent human–nature relations exists among hunters in rural Vermont and that some of the contemporary meanings of hunting are directly related to the conditions of modern life. I did not, however, explain how interdependent human–nature relations actually relate to specific hunting practices. As a result, a number of important questions—questions that nonhunters often ask—are left unanswered, such as "Do hunters just like to kill?" and "Are there any shared ethics involved with hunting?" In this chapter I offer detailed descriptions of hunters' feelings about acceptable hunting practices, the satisfactions of "the chase," the ways that hunters treat slain animals, and the importance of eating game.

Sport

I generally refrain from using the common term *sport hunting* because I fear that equating hunting with sport belittles the deep meaning it holds for many people. The term *sport*, as used by the hunters I met in Vermont, simply did not correlate with popular uses of the term that relate to organized games or lighthearted recreation.[1] Hunting was more often described to me in terms of *satisfaction* and *meaning*. Even the hunters who regularly played sports such as golf or softball made it very clear that hunting was a vastly different endeavor because it dealt with serious existential dilemmas in ways that traditional sports cannot. This is not to say that enjoyment was not a part of hunting. Unless it was necessary for survival, people wouldn't hunt if they didn't enjoy doing it. The point here is to realize that the enjoyment of hunting is complex, and it is seldom measured exclusively by the animals one has killed.

Terms like *sportsman* and *sportsmanship* were often used by hunters, and they conveyed meanings that were similar to, but certainly more serious than, those they would have for, say, the typical color commentator during an athletic event on television. This use of the word follows the second definition offered by the *American Heritage Dictionary* (1993): "Conduct and attitude viewed as befitting athletes, as fair play." Proper conduct and attitude and a concern for fairness were generally considered very important to hunters who discussed their hunting techniques with me, and I routinely heard criticism of hunters who did not follow game laws, were unsafe in the field, or shot animals before they were definitively identified.

One very common way that the sporting aspect of hunting was described was in relation to the difficulty of the hunt. To most hunters, a particularly easy hunt was not highly desired. This is not to say that hunters were not thrilled when they killed a deer fifteen minutes into the opening day of hunting season; rather, what they are referring to is described by the old saying "It's like shooting fish in a barrel." If, for some reason, the hunt does not give the animal a chance to survive, then it would not be considered sporting. For example, a contractor in Pittsford who only bow-hunts told me that, in addition to just being a little scared by all the people running around with high-powered rifles, he dislikes firearm hunting because he feels it is too easy. He said, "If you don't have to work for it, then there's no sport in it." Clearly, in this statement the term *sport* carries a strong ethical connotation.

A woman in Salisbury elaborated on this issue and touched more directly on the relationship of ethics and sport: "I don't know, I just keep thinking of the sport issue. It was sport for me because it's not like I have a hundred acres of land with thirty-foot-high fence and I've got this animal in there and I'm partaking unto his little area where I'm gonna end up shooting it. I'm on his turf, and his terms, and I've gotta try to outsmart him. Now, how do you do that when he's got the full advantage? I'm just a dumb little shit out here, you know—with time on my hands."

Ironically, I happened to meet a man in early 1998 who actually was placed in a situation somewhat similar to the hypothetical one described above. Locally recognized as a very successful hunter, he was asked by a hunting video producer to literally "play" the role of a deer hunter in a very controlled situation that was being presented as an actual hunt. He said that during and after the filming he felt bad and that he still regretted doing it. As opposed to a hunt, he said it felt like an "assassination."

The idea of reducing the practice of hunting to the act of killing an animal was disturbing to most hunters I met in Vermont. They realized, quite rightly, that this is the kind of simplification often engaged in by antihunters. A point of critical importance is that, according to most of the hunters I dealt with, the satisfaction of hunting is derived from *engaging in the process* of hunting; not simply the act of killing of an animal. This line of reasoning actually agrees with any dictionary definition of *hunt*, which refers to the *pursuit* of animals. Hunters who simply found it "fun" to go out and shoot deer were rare, and they were considered to be out of the ordinary by most other hunters. As one man said, "That's what I wish a lot more people would understand. It's not just killing animals. It's a lot more than that. . . . The problem—what a lot of people don't realize is that most hunters care an awful lot about the animals that they take. . . . I would say that probably 95 percent of the hunters feel the way I do." These sentiments are reminiscent of José Ortega y Gasset's famous claim that "one kills in order to have hunted" (1985, 97). Although among rural Vermonters, as I will explain, it might be more accurate to say that one *eats* in order to have hunted.

A tradesman in Rochester (who would later invite me to his deer camp in the Northeast Kingdom) said, "You know, to me you don't have to kill an animal to enjoy a hunt." He also took great pride in pointing out that his uncle does not kill the bears that he pursues with his dogs. "That's a hunter to me. You don't need to kill 'em." He went on to tell me a personal story that serves as an excellent example of the importance of hunting ethics, as well as the distinction between hunting and killing. One summer he noticed that a particularly large buck was feeding at a local orchard every day in the late afternoon, and one day he decided to go over to the orchard and shoot the deer, even though hunting season was still months away. "I laid there 'til dark and watched him, but I said 'I can't shoot that deer.' I got more enjoyment out of just watching him. Everybody had seen him, you know, and I wanted that big set of horns." I asked him if it really would have mattered if he had gone ahead and shot the deer. "Yeah, it would to me. I didn't think so at the time, but when I got up there and started watching him I didn't have the heart to shoot him." Some hunters, he said, would have shot the deer. He mentioned one local man in particular who has quite a reputation as a poacher. "He'll just kill 'em to see 'em drop."

Drawing this distinction between *hunting* and *killing* was not unusual for the hunters I met. For example, another man from Rochester said of out-of-state hunters, "They'd rather kill than hunt." A man in Middlebury echoed this sentiment when he remarked that out-of-staters are "just out to kill something."

It's important to note that these critiques of a perceived overemphasis on the kill alone were sometimes directed at other members of the local community as well. Some people who generally adhered to the negative stereotype of hunting that is broadcast by antihunting groups (such as indiscriminate shooting, illegal killing, or failing to retrieve wounded animals) were basically abhorred by everybody I met. Unfortunately, I did not interview (or did not realize I had interviewed) anybody who fell into this "outlaw" category. The other type of hunter sometimes criticized for flawed hunting ethics is the trophy hunter. I interviewed a fair number of trophy hunters; most were exclusively trophy deer hunters, though a few had killed mountain lions, wild boar, and certain African species. The trophy deer hunters placed a high premium on getting a buck with large antlers and would often travel to places like Saskatchewan in pursuit of deer that might have a world-class rack. Several people I encountered openly disapproved of this kind of hunting, speaking of it in terms similar to those of the man who told me out-of-staters hunt "just to kill something."

The best available evidence to corroborate the claim that hunters in Vermont do not hunt for the joy of killing animals is that most hunters are not very successful. During the 2000 deer season, for example, an estimated 83 percent of Vermont hunters did not kill a deer (VFWD 2000, 15). Still, they continue to hunt year after year because they find it so enjoyable. An excellent example of this comes from an interview with a man from Middlebury who was over eighty-seven years old at the time of our interview. He told me that he hunted for twenty-five years before he got a deer! During those early years, he pointed out, he was usually a "driver"—a person who walks around in the woods in an effort to drive deer to people "on stand" who are strategically positioned to have a clear shot at the moving deer. Not surprisingly, drivers generally have less chance of getting a deer than somebody on stand. When I asked him whether or not he cared about getting a deer all those years he replied, "I didn't care if I got one or not. I was interested in being out in the woods and in hunting." In similar fashion, a man in East Middlebury said, "I went twenty years without getting a deer and kept hunting because I enjoyed the hunting."

The woman in Salisbury I quoted earlier on the sporting aspect of hunting spoke of her enjoyment of what some might call "unsuccessful" hunts: "I love to hunt because I just like to go out there and I could care [less]—I'm full. I don't need to eat deer. . . . If I added up all the days and hours I've put into hunting and how many animals I got it certainly isn't gonna look very good—not for me anyway."

Many other men and women made comments similar to the ones I have quoted. The obvious question, then, is, What exactly constitutes hunting for these hunters in rural Vermont? I asked this question directly during many of my interviews and found that activities included under the conceptual umbrella of hunting went well beyond the actual time spent in pursuit of game. Hunting was always defined as inclusive of preseason scouting and often of eating the game after the season's end.

Clearly, the satisfactions of hunting are derived from more than the isolated act of killing an animal. I also want to make clear that they are not even necessarily involved with the preparation or anticipation of the kill. Nonhunters might be surprised by just how much enjoyment many hunters derived from simply being in the woods. Because hunters often spend a great deal of quiet time outdoors, they have the opportunity to witness nature in a way that few people ever do, and these incidents—a chickadee or chipmunk hopping on one's shoulder, seeing an eagle attacking its prey—are highly valued. A ninety-year-old man in Rochester told me, "I hunt for the pleasure of being out in the woods"—even though he had once been charged by an angry black bear.

A question that came to my mind during my fieldwork, and that a number of nonhunters have asked me as well, is, How often do these hunters ever go out in the woods without the intention of hunting? Understandably, all the claims of loving to hunt because it gets them outside would be undercut if hunters only went into the woods when they had a chance to kill something. Consequently, the answer to this question has important implications for my work. What I found was that most of the hunters were outdoor-lovers in general. Just as they were very likely to spend time in the woods hunting for grouse, rabbits, or deer during the fall, there was also an excellent chance that they would be off snowshoeing, skiing, riding horses, rock climbing, camping, fishing, or picnicking at other times of the year.

A statement that I commonly heard from hunters was "Hunting gives me an excuse to get out in the woods." Initially, I was a bit cynical about this remark. Why, I wondered, do you need an *excuse* to go out in the woods, anyway? And why do you have to go out in the woods with a gun to have an enjoyable outdoor experience?" As time went on, however, I began to understand what they meant. After all, how many people could get away with telling their husbands or wives that they are planning on taking long walks in the woods every day for the next two weeks, or that they will be staying in a cabin in the mountains with their friends for a couple of weeks, if it were *not* for hunting?

Simply put, in Vermont hunting is not considered to be a frivolous way to spend one's time. It is so deeply entrenched in the rural lifestyle that to deny someone, especially a man, the chance to hunt would be considered extremely questionable behavior. With this in mind, we can see that hunting season is often a way for someone to take a vacation from work and essentially relax without being thought of as "wasting time." As I noted in the previous chapter, hunting is a form of *productive leisure* that, through its potential of material production and its historical link to a frontier past, is thought by many people to be much more worthwhile than lying on a beach.

One stockbroker—a native Vermonter—provided some particularly enlightening commentary on this issue of hunting's "getting you outside." While describing all the amazing things he had seen in the woods while hunting (by the way, he has only shot a couple of deer in the last twenty years), he mentioned that he loves bird-watching. Hunting season, he explained, gave him a viable way of watching birds, since he was not about to join a bird-watching club and, as he described it, be expected to wear a little khaki vest and run around with binoculars dangling from his neck. That just wouldn't cut it for this former college football player. But peering at songbirds through the scope of his high-powered rifle while hoping for a deer to show up was one of his life's true pleasures.

The Chase

Whether in theory or in practice, hunting requires the pursuit of animals. While there is no doubt that many hunters do not expect to make a kill every time they go hunting, the key aspect that transforms a walk in the woods into hunting, and also provides enough interest to keep people hunting every year, is the pursuit of animals. Hunting is generally described as an exciting activity that requires intense mental focus and often demanding physical exertion.[2]

Of all the aspects of hunting that make it such an engaging pursuit for many men and women, the most universal is probably the intellectual challenge of trying to figure out where the game is and how to put yourself in position for a good shot. A good example of this comes from a construction foreman who took a hunting trip to Maine for the chance to kill a trophy buck. "I didn't get that deer, but I was so happy—so proud of myself—that I was able to get that close to a trophy buck. That is probably—that's one of the most memorable instances in my life that ever happened to me in hunting, and actually in my life really."

After getting to know so many hunters and personally hunting in Vermont for several seasons, I gained a great respect for the abilities of people who were consistently successful, particularly those who were expert deer hunters. Since then, I have surprised more than one nonhunter who asked what I think of hunters by responding, "Well, the good ones are really smart." Particularly in Vermont, where the habitat is far from ideal for the production of large deer herds, deer hunting is a difficult undertaking. As I have mentioned, many of the hunters I met in Vermont had experienced fifteen or twenty unsuccessful hunting seasons—sometimes consecutively—by the time they were forty years old. Granted, most of the people who fall into this category probably did not work as hard at hunting as those who were very successful, but they clearly knew enough about the outdoors to be successful hunters in other states that have larger and more accessible deer herds.

The difficulty of hunting deer in Vermont is something that most suburban people I meet—particularly those who have never hunted—have a very difficult time imagining. Many of these people can list several places they pass every day on their way to work where they can count on seeing large herds of grazing deer. The situation in Vermont is much different. In eighteen months in Addison County, for example, I *never* saw a herd of deer as large as those I can regularly see in the cornfields behind my mother's farm in western New Jersey. Certainly the difficulty of killing a deer in Vermont is one of the reasons why it can be such a exciting occasion. On the other hand, this difficulty sometimes leads to endless assaults on Vermont's Department of Fish and Wildlife for its supposed mismanagement of the herd.

To provide an example many of us can relate to, think of how seldom people hiking in the mountains actually see large wild animals. They might see squirrels and chipmunks and the occasional porcupine, but how often do they see something on the order of a deer? Not very often, and when hikers do spot a deer it is usually considered the highlight of the day. Now take this a step further and think about how many of those few hikers who have ever seen a deer actually know *why* they saw that deer. The number becomes even smaller. There is a great difference between stumbling on a deer on a walk and successfully *predicting* where that deer will be at a particular time on a particular day.

After my experience in Vermont, I was not surprised to learn that Stephen Kellert of Yale's School of Forestry and Environmental Studies found that certain groups of hunters were more knowledgeable about the outdoors than any other group of people that he surveyed (Kellert 1976).[3] Hunters must have a sound knowledge of the overall ecology of an area and the particular behavioral tendencies of the animal they are seeking. What is the acorn crop like this year? What about the beechnuts? What are the farmers growing? Are there any abandoned orchards in the woods? Where are the likely waters sources for deer? Where do the prevailing winds come from? What are the travel routes for deer in this area, and where do they bed down during the day? If spooked, as they often are during the deer season, what are the routes of dispersal and where do they lead?

Besides the cerebral aspect of pursuing animals, the other source of enjoyment that is often mentioned in association with the chase is the exploratory adventure of going new places in the outdoors. Many hunters reported that they were more likely to venture into unknown territory when hunting than when they were simply out for a hike. They also anticipated the coming of deer season as a time when they would have the opportunity to ramble through the woods for several days on end. One man, for example, told me that during hunting season he looks for new places to go for picnics with his family when the weather warms up.

Successful hunters also know how to track, use weapons, field-dress animals (remove the entrails), and generally orient themselves in the woods. Clearly, hunting demands a vast amount of expertise in a variety of different domains. As one man put it: "A person who gets a deer every ten years is lucky. A person who gets a deer every year—there's more to it than luck."

A question that may come to mind is why the experience of being in the woods seems to be enhanced during the hunting season. There are, after all, plenty of people who find great pleasure in hiking and never even consider going hunting. But the question itself assumes that hunting is a morally dubious endeavor and, as such, makes more sense to people who are ambivalent about hunting than those who participate in hunting. For instance, would anybody ever ask a mountain biker why they have to ride a bike to enjoy a day in the outdoors? The broad and simple answer to the hunting/hiking question is that hiking does not possess the cultural significance of hunting to many Vermonters. To many it feels less engaged as well—watching the world rather than participating in it. Another, perhaps less intuitive answer I was given is that hunters achieve a degree of mental focus and heightened awareness of their surroundings that they simply cannot attain in a nonhunting situation. Being in the woods, therefore, becomes more interesting, more exciting, and more meaningful when hunting.[4] As one man in Ferrisburgh commented, "It just feels right."

Eating Game

Of all the opinions I heard on the practice of hunting that, together, make hunting in rural Vermont an ethically regulated process, probably the most prevalent was that slain animals must be eaten. An animal that is eaten is an animal that has not been wasted. As one woman put it: "I don't have a problem with—like I said—killing just about anything as long as something's gonna eat it. But it kills me—it breaks my heart to see something dead in the road."

The disdain that hunters expressed about not eating animals does not come as a surprise from the perspective of a discourse of interdependent human–nature relations. In its capacity as a natural resource, the nonhuman world's material relationship with humans hinges on exploitation; in the case of game animals, human consumption becomes the legitimate use of animals such as rabbits, grouse, and deer. Killing a game animal for any other reason is undeniably viewed with suspicion. This is precisely why certain Vermont hunters I know have little respect for people who trophy hunt. The perception is that trophy-hunters are leaving the crucial aspect of subsistence out of the hunt. Yet I did not speak to a single trophy hunter who would condone killing an animal if it was not going to be eaten, even if it was eaten by the hunting guides and not the actual hunter.[5]

As I note elsewhere, a number of hunters were concerned that many Americans had effectively lost touch with the material reality of their existence. Killing and eating animals, these hunters claimed, prevents them falling prey to a similar fate. More specifically, some hunters complained of the hypocrisy of antihunters who choose to eat meat. Many thought that eating a farm animal that somebody else was paid to kill was far more suspect from an ethical perspective than taking personal responsibility for killing a wild animal. One hunter in East Middlebury who grew up on a farm, and who said he was always turned off by the slaughter of farm animals, remarked, "I think there would be a lot of vegetarians if people ever went to a slaughterhouse."

Gestures of Respect

It was common during my research in Vermont to be told of certain gestures of respect that were carried out after an animal was killed. In one way or another, these gestures are meant to offer thanks or honor to the slain animal. For example, a man in East Middlebury told me, "I usually—after, when I bag one—I actually do like the Indians used to and that's to say a prayer for them." The woman I mentioned in the previous section, who always cries when she kills a deer, said, "And I always do a little thing. I sort of thank them for giving their life to feed us and that's just what they're doing." One man told me that he had recently held a deer's muzzle and inhaled its last breath as it died.

Probably the most elaborate gesture that I was told about came from a man I met at a camp in Ripton. The day I arrived at camp, I noticed he was wearing a long necklace with a deerskin pouch hanging from it, and I wondered what it was all about. The next night, as we sat next to each other at dinner, he explained that in the pouch are the ashes of his dead father. When he kills a deer—in addition to offering a prayer of thanks—he sprinkles some of the ashes on the "gut-pile" that he has removed from the animal. Since he knows that this viscera will be eaten by other animals, this act insures that his father will be eternally incorporated into what he calls the "real world."

A more typical way of showing respect to an animal is having it taxidermically mounted. Usually only the heads are mounted, but occasionally people have the entire body done. Most dramatically, one woman had a stuffed mountain lion displayed in a glass case in the middle of her house, and one man had a bear on display—minus the glass case. One man I talked to in Bristol said that he mounted animals so they will not be forgotten and to honor them; a woman in Lincoln said she did it to show respect and thanks to the animal, to relive the experience, and to "bring nature in here." And the man in East Middlebury who said he prays for the animals he kills called the deer head on his wall as a "visual reminder." He also said, "This may sound a little weird, but when we are eating venison I actually salute to him—thank you."

Before I came to know these hunters, I was very skeptical of almost any kind of taxidermy; it certainly never occurred to me that mounting an animal's head on the wall might be done out of respect. The only reason I could think of for mounting an animal was to show it off to your friends. While I certainly would not claim that all works of taxidermy originate from a reverence for the slain animal, what I learned in Vermont has changed my thinking about the possible meanings of this common practice.

Following the theme introduced in the previous chapter, gestures of respect toward slain animals are predicated on a worldview that considers killing to be a natural part of earthly existence for many animals—including humans. This does not mean it isn't a complicated matter. Along with all that killing animals provides (food, self-esteem, an enduring connection with physical surroundings, a feeling of environmental health) it also brings sorrow for the deaths of beautiful, intelligent creatures. There are probably many ways that people could deal with this situation. The most common way among the hunters I know in Vermont is to show respect toward the animals that are hunted and to be thankful for their lives when they are taken.

5

Gender Transformations

Considering the extreme male domination of hunting in Vermont, it is not surprising that males and females often take distinctly different paths to hunting. Likewise, the personal experiences of male and female hunters can differ greatly as well, with men living out a rural cultural ideal and women struggling against this ideal. In this chapter I will describe the social production of male and female hunters and discuss the various pressures that female hunters endure, and I will ultimately argue that female hunters are engaging in authentic acts of political resistance when they hunt.

My analysis of male hunters seeks to determine whether hunting in rural Vermont is generally a hypermasculine, perhaps even misogynistic, cultural production. I take into consideration male hunters' notions of masculinity, the contours of masculinity in rural Vermont, male hunters' opinions of female hunters, and life at all-male deer camps in the Green Mountains. Based on this data, I conclude that male hunting in rural Vermont, though clearly informed by traditional expectations of American masculinity, is not primarily an expression of hypermasculinity that degrades females in an attempt to define males.

In many respects, this chapter amounts to a snapshot of social change in the making. Consequently, it reverberates with a sense of personal and social indeterminacy. Apparent contradictions and unanticipated contingencies abound when talking to rural Vermonters about gender and hunting. The increase in the number of female hunters is both the consequence and the cause of shifts in prevailing gender ideologies, and, as we will see, both male and female Vermont hunters are moving through this transitional time in ways that are often unpredictable.

Beginnings

Fathers act as gatekeepers, teaching hunting to their sons but not to their daughters. The rare female hunter is often the first-born or only child of an avid hunter. Females may also learn from boyfriends and husbands in order to share this recreational activity. The woman who is taught to hunt by another female in the absence of a male hunter is most unusual, and females raised in urban areas with nonhunting fathers almost never become hunters themselves. All of these characteristics highlight the critical role of social learning in determining who becomes a hunter.

THOMAS HEBERLEIN, "Stalking the Predator: A Profile of the American Hunter" (1987).

Mirroring the general national hunting profile, hunting in Vermont is characterized by a dramatic contrast in male and female participation. Nationally, 91 percent of all hunters are male and 9 percent are female; in Vermont the figures are 92 percent male and 8 percent female. But these nearly equivalent statistics actually obscure significant differences in the social character of hunting between Vermont and the nation as a whole, because a relatively large portion—at least by today's standards—of Vermont males hunt: 21 percent. In comparison, only 10 percent of all males hunt nationally. Only a very small percentage of women hunt: 1 percent of all American women and roughly 2 percent of all Vermont women (all figures from USDI 2006a, 29, and USDI 2006b, 25).[1]

These statistics reflect the personal experiences of male and female hunters in the United States. Young girls and boys usually have greatly differing relationships to hunting. For a young boy in a rural area, for example, being interested in hunting is fairly ordinary. Boys are often encouraged to hunt, and if they eventually end up hunting on a regular basis, they usually begin in their youth. For males, then, hunting could be said to be a performance of rural masculinity; or, borrowing a term from the anthropologist Sherry Ortner, hunting is one important way that rural males "do" masculinity. A girl, on the other hand, may have never known, or even seen, a woman hunter, and she will almost certainly not be expected to be interested in hunting; as a matter of fact, one woman I interviewed, when asked about her reasons for hunting, immediately said, "Because it's not expected." Consequently, if a female does begin to hunt it is most often as an adult (Jackson and Norton 1980; Heberlein 1987).

My research in Vermont generally agreed with these sociological profiles. The great majority of the male hunters I met were introduced to hunting as boys and continued hunting uninterrupted into adulthood.

Most females began hunting as adults. Moreover, several of the female hunters I interviewed were raised by their fathers in the absence of a mother or were daughters of men who had no sons, and nearly half of the twenty-three women I formally interviewed were introduced to hunting by a boyfriend or husband—a process sometimes referred to as "courtship hunting." For women who are not opposed to the idea of killing wild animals, but who were never given the opportunity to hunt when they were younger, courtship hunting has long been their best chance of ever learning to hunt. One Middlebury woman I interviewed went woodchuck-hunting on her first date with the man she eventually married. She hunted for the rest of her life, nearly fifty years, and was quick to point out when we spoke that she could never have done so much hunting without "a great babysitter." Like most women all around the world, women in rural Vermont bear the heaviest share of child-care responsibilities. As a result, their opportunities to hunt often depend on reliable child-care arrangements. I have never heard of a man being unable to hunt because he had to baby-sit his children.

As Heberlein notes (1987, 8, quoted in the epigraph to this section), women often become involved in hunting because it provides an opportunity to share a recreational activity with their husbands. Indeed, several women told me that this desire was the root of their interest in hunting. For some women, in fact, this time spent hunting with their husbands is more valuable than one might expect, because the husband might be the kind of hunter who spends all his free time in the woods from October until December. As one avid bow-hunter who takes off much of the fall and winter to hunt told me, "If she doesn't hunt, she doesn't see me." His wife, the holder of a local political office at the time we spoke, had also become an enthusiastic bow-hunter.

A woman from Bethel (in nearby Windsor County) said she initially got involved in hunting "because he was doing it." As a self-described housewife who enjoyed the chance to get away from everyday domestic duties, she thought hunting with her husband would be an enjoyable diversion. "I'm an outdoors person, so the hunting just kind of came with the husband little by little." She soon realized why her husband found it so satisfying; as she commented to me, "I absolutely love it." This woman has three grown daughters, two of whom hunt. One hunts only birds; the other killed a bear with a bow in Nova Scotia. The third daughter is an ex-vegetarian who now eats only organic food, including wild game killed by her family members.

As one avid and successful female deer hunter explained, getting your husband to take you hunting is sometimes easier said than done. This woman, who was in her mid-forties and on the board of the local Humane Society when I met her during my initial fieldwork, recalled the early days of her marriage and her husband's initial reaction to her desire to hunt: "The first year was like, I watched him go off to deer camp and I set there, and I was two weeks all by myself thinking this is bullshit. . . . If I ever stick to this guy, you know, I'm gonna have this two weeks every year all by myself and I thought no, I'm not gonna! . . . And I remember the day I said something to him about—Why don't you teach me how to hunt? And he just kind of looked at me and smiled because that wasn't the way that he was brought up. You know, all the guys got together. It was a macho thing." Her husband refused to take her hunting, proceeded to go to deer camp as planned, and advised his young wife to find somebody else to take her hunting. To his surprise, she did just that, enlisting her cousin's husband as her personal hunting instructor. A few days later, however, her husband came home unexpectedly from deer camp. As she describes it, when he came in the house he simply said, "You wanna go hunting?" They have hunted together ever since—over twenty years—and he has never gone back to deer camp. And they wouldn't have it any other way. The only negative aspect of this arrangement is that her husband has been the object of some kidding by his friends. As she said, "They look at him, you know, like he's pussy-whooped or whatever they call it, you know."

The desire to spend more "quality time" with their husbands was certainly not the only reason why women began hunting after marriage. Some simply had a strong personal interest in hunting, and their husbands were happy to teach them. Nearly all of the courtship hunters shared a great affection for hunting—these were clearly not women who were merely hunting to be with their husbands, even if that may have been the original reason for their involvement. Hunting was now an important part of their own lives. The woman I quoted in the previous paragraph, for example, once told me: "I just love it. I just love it. I look forward to it. I'm like a little kid. . . . I don't even sleep. I just lay there. . . . I see 'em and I just get myself so—it's like [makes panting noises] you know, and you're kinda sittin' there and the sweat starts and here it is twenty below zero."

As we have seen, most hunters—male or female—find hunting to be exciting, challenging (both mentally and physically), adventurous, and deeply meaningful. The critical difference, as Heberlein pointed out, is that young females are far less likely to be exposed to hunting than males. But for those women who do not already harbor negative feelings toward the idea of killing wild animals, hunting is often an activity that they soon grow very fond of once they begin.

Although it is doubtful that any researcher would question the general validity of Heberlein's characterization of the social production of hunters, social changes have occurred in the ensuing years that make certain aspects of this characterization less dependable than they were in 1987. I am referring specifically to the role of males in the production of hunters. It certainly appears, at least in the central Champlain Valley of Vermont, that the importance of males in the dissemination of hunting knowledge is decreasing.

In Vermont, there are two main sources of this change. First, we must take into consideration the fact that, in recent years, women are accounting for a growing proportion of the American hunting population. While the percentage of women who participate in hunting has remained steady since the early 1990s at about 1 percent, male participation rates have decreased significantly—from 14 percent in 1991 to 10 percent in 2006. Today, then, about 9 percent of American hunters are women, as opposed to about 8 percent in 1991 (USDI 2006a, 29; USDI 1993, 36). This increasing prominence of female hunters, as many women have told me, mirrors the widespread involvement of women in other male-dominated facets of American society. One Vermont female hunter told me, "More and more women are doing men's things"; another said today's women are "more their own person." Another located the source of these changes in the schools: "I think in the school systems young females are being taught that you can do—be anything you want to be. You don't have to be a nurse or a secretary. You don't have to get married out of high school and start a family. You can do anything that a man can do and I think it's starting to work."

The first wave in the rise of female hunters, then, occurred among women who were influenced by the women's movement that was sweeping the entire country when they were coming of age in the late 1960s and 1970s. These pioneering women were not taught to hunt as young girls, but rather began under their own impetus when they were older. They have, in turn, greatly increased the chances of today's girls learning to hunt even before they enter high school.

Although there are not yet reliable statistics on the number of girls hunting in Vermont, careful observation suggests that they are on the rise. For example, the people who sell the most licenses in the Middlebury area—employees at sporting goods stores—report that they have been seeing far more young girls buying licenses in recent years than in the past. Anecdotal evidence of the increase of young female hunters is also positive. For example, in 2000 I spoke to a Middlebury woman who told me that she had taken her young granddaughter turkey hunting for the first time. And I once saw a photograph of young girl with her first deer on the wall of a deer camp in Ripton—the granddaughter of the camp's owner. It is almost inconceivable that this would have happened in the 1970s or 1980s.

As a result of the increased participation of women in hunting, there are more women and girls who are being taught to hunt by women. This is a dramatic change from the 1960s and 1970s. Today, women in Vermont are taking a special pride in teaching younger females to hunt. As one woman from New Haven Mills, who was probably in her late twenties, said, "I'm trying to be a role model for my nieces." She knew from personal experience that girls are often left out of hunting even when they are interested. Although she took the required hunter-safety class when she was only nine years old, she was never asked or encouraged by her father to join him when he went hunting, even though he was the instructor of her hunter-safety course. Consequently, it was not until she was eighteen that she began hunting with her boyfriend.

In the coming years, I would not be surprised if there is more pronounced increase in the number of female hunters than than what has occurred over the last twenty years, because it will be the first time in recent Euro-American history that so many mothers will be hunting with daughters, aunts hunting with nieces, and grandmothers hunting with granddaughters. The people I've talked to in Vermont say that there are more twelve-year-old girls hunting in Vermont today they can remember. Nevertheless, we should not assume that hunting in rural Vermont is now a sexually egalitarian activity: I've never heard a man say that he hunts in order to spend time with a wife he would otherwise not see during hunting season, and boys still are far more likely to be encouraged to hunt than girls. An excellent example of the persistent tendency to consider hunting a singularly male endeavor came out in a conversation I had in 2004 with a hunting grandmother. She said that her daughter-in-law has proclaimed that she will not allow her daughter (whom the grandmother hopes to "see with a gun in her hand" someday) to hunt, but that her grandson, who happens to named Hunter, is being actively encouraged to hunt.

I also know women who are teaching, or have taught, boys to hunt. It is somewhat remarkable that in Vermont—a state where nearly a quarter of all men hunt—there are now boys learning to hunt from women. Indeed, state governments and organizations concerned with maintaining the popularity of hunting are now focusing on teaching girls to hunt. This strategy hinges on the belief that, in a society with so many single mothers, female mentors will play a fundamental role in maintaining the popularity of hunting for years to come.

A good example of this relatively recent interest in the recruitment of girls into hunting is the official Vermont Fish and Wildlife Department calendar for 1997. Designed as a celebration of the hundredth anniversary of the state's deer hunting season, it displays only one picture: a young girl hunting with a middle-aged man (presumably her father). The agenda behind the design of this calendar could not be more transparent.

The second primary reason why women are learning to hunt in the absence of men is the emergence of two innovative programs: Becoming an Outdoors Woman (BOW) and Vermont Outdoors Woman. BOW was originally developed by Christine Thomas (a professor at the University of Wisconsin–Stevens Point College of Natural Resources), and seeks to teach women a wide variety of outdoor skills that they otherwise might not have the opportunity to learn.[2] As Thomas stated in a 1991 *Milwaukee Journal* article, "One reason why women don't really take part in hunting and fishing is because as children no one taught them the skills to make their participation more enjoyable. It's not much fun if you don't know what you're doing" (quoted in Vanden Brook 1991).

Generally, BOW seminars (which have been available through the Vermont Department of Fish and Wildlife since 1995) depend on the assistance of local outdoor experts (in anything from hunting and fishing to camping and orienteering) to run the various workshops that take place at each regional seminar. These instructors are often women, but many are men as well. According to women I have spoken to in Vermont and other states, they generally feel more comfortable with female instructors. There is also a fairly general feeling that the average man makes things far too competitive to be enjoyable. As one woman joked, "I know if a bunch of men were to do a Becoming an Outdoors Man there would be prizes!"

Vermont Outdoors Woman (a division of the Vermont Outdoor Guides Association) began in 2001 and aims to "encourage and enhance the participation of women of all ages and abilities in outdoor activities, through hands-on education."[3] Vermont Outdoors Woman is perhaps best known for its annual summer, fall, and winter Doe Camps that offer instruction in a wide variety of outdoor activities and skills, including, among others, dog sledding, ice-fishing, fire-making, archery, firearms safety, kayaking, fishing, self-defense, animal tracking, and chain-saw safety. Vermont women can also participate in the National Rifle Association's Women on Target program, which is intended to teach women "shooting and hunting skills in a safe and supportive women-only atmosphere."[4]

Ideology and Female Resistance

However common hunting by females may be cross-culturally (Brightman 1996), in the United States it is an exceptional occurrence. Despite the recent shifts in the sexual composition of the hunting population, female hunters still represent only a small minority. Nevertheless, male and female hunters in the Addison County area share a great deal in common. If one were to focus exclusively on hunting techniques, the kinds of satisfactions found in hunting, and issues of human–nature relations, very little difference between male and female hunters would be apparent. But because these women—and girls—are running headlong into some deeply entrenched American gender expectations, the personal hunting-related experiences of most female hunters I know differ in many respects from those of male hunters.

Male Hunters' Perspectives

It came as something of a surprise to me that the male hunters I interviewed in Vermont made very few disparaging remarks about women who hunt. Several of them, in fact, were extremely complimentary of female hunters. As my research progressed, however, I was no longer surprised when a man would say something like "I'd say women are better hunters." The middle-aged manufacturing plant worker who said this went on to explain that women had better depth perception and were less likely to be color-blind. Another man went even further: "Every woman hunter I've ever seen is probably a better hunter than I am. . . . Women will do exactly what you recommend to them. I don't know how the hell they do it." A archery-safety instructor I spoke to in 2004 told me that 80 percent of a recent class of about twenty-five students were women, and that they were very eager to learn and generally better listeners than his male students. These compliments on the ability of women to listen touch a common theme that arose in conversations with both male and female hunters: the idea that women are more patient than men. And this is no trifling matter, since it is a widespread belief among hunters that patience is one of the most important attributes a hunter can possess.[5]

A woman in Ripton touched on this theme when she recalled that when she began hunting she followed every directive men gave her about how to hunt; spending hours on end sitting in deer stands on bitter cold days, moving as little as possible for fear of startling approaching deer. It was only later that she realized that what these guys were telling her often had very little to do with how they were actually hunting themselves. It seemed to her that the men were incapable of just sitting still and not unwrapping a candy bar, digging into their packed lunches at nine in the morning, or even urinating out of their tree stands. As she put it, "Men always seem to have a problem." At the time we spoke she was a serious hunter with some bona-fide trophy deer to her credit, and she thought back with amusement, and perhaps a touch of irritation, at having believed that these guys actually knew what they were talking about.

All of this is not to say, of course, that female hunters do not receive any negative treatment from men, including men who hunt; as we will see, they certainly do. My point here is simply that it would be wrong to assume that male hunters generally attempt to block the entry of women into their special domain just as men have tried to block the entry of women into any number of male-dominated aspects of American life. There certainly was not a widespread disapproval of women becoming involved in hunting among the male hunters I encountered in Vermont, whether because men enjoy the opportunity to share quality time with their wives, the satisfaction of knowing their wives or girlfriends can relate to something so important to their lives, or simply from the comfort of knowing that women (the most likely people to be animal rights activists, incidentally) are becoming more accepting of hunting. This lack of concern about female participation in hunting does not support the contention that hunting is an exclusive male bastion utilized for the performance of hypermasculinity.

Suspicious Men

Despite the amicable attitudes expressed by the men I spoke with in Vermont, female hunters are surely viewed with skepticism by certain male hunters. A prime example of the doubts female hunters evoke in some men are the suspicious murmurs that can be heard at the state-mandated

deer checking stations where hunters are required to bring their kills. The
large sporting goods store I worked at serves as a major checking station, so I witnessed many such incidents. While a twelve-year-old boy will
be congratulated on his kill, a thirty-year-old women might be doubted
by onlookers. According to women I spoke with, this skepticism can be
quite biting at times, with men making sarcastic remarks along the lines
of "Nice deer, Jim" that are directed toward the woman's husband as
they place the deer she killed on the biologist's scale. The implication of
such a remark, of course, is that the couple is simply pretending that the
woman killed it to allow her husband to kill more deer than he is legally
allowed. It must be kept in mind, however, that there is a long-established
practice among rural Vermont families of buying a hunting license for
every family member and "filling the tags" any way they can.[6] According
to many people, it was once quite common for farm women to bring deer
to checking stations that they clearly had no hand in killing.[7] Undoubtedly, this historical legacy fuels at least some of the skepticism expressed
toward female hunters at the state's checking stations.

Female hunters confront other kinds of harassment as well. One woman told me that she was regularly teased about her hunting by men at
work, who obviously doubted her skill. Ironically, this particular woman
has been hunting since she was a little girl and probably knew more about
it than many of her male coworkers. Another woman recalled that two
men had followed her into the woods one day when she was hunting, in
what she assumed was an attempt to take advantage of her careful scouting. Still another, rather common, complaint of women is that men treat
them as if they are children who have never hunted before. This results
in constant reminders about very basic things, such as not pointing their
guns at other people.

The net result of this skepticism is an environment that puts a great
deal of pressure on women to be successful hunters; as well as a double
standard with regard to any mishap that may occur. For example, it is not
uncommon for men to become disoriented and lost in the Green Mountains while they are hunting. Very often, a fairly dramatic story comes
out of these ordeals and the adventure aspect of hunting is reconfirmed.
I have one of these stories, and many more experienced Vermont hunters
have them as well. If a woman were to get lost, however, I have no doubt
that certain men would turn the story into a critical commentary on the
ineptitude of female hunters. The same would apply to women missing
open shots, inflicting nonmortal wounds, losing the track of wounded
animals, or even getting their trucks stuck in the mud.[8]

Many female hunters told me that they provoke negative reactions from women as well as men. In fact, some female hunters viewed nonhunting women as a greater irritant than men. As one woman said, "I think I get more of the odd reactions from women." Another woman reported that women have asked her if she is "trying to prove something" by engaging in such a traditionally male activity. Several also claimed to prefer male companionship to female companionship, sometimes even citing supposed female personality characteristics such as "bitchiness." There appear to be two primary reasons why some female hunters are bothered more by comments from women than those from men: whereas men may critique the general idea of women hunting, nonhunting women more often critique hunting itself, expressing ethical reservations about the entire enterprise; and because so few women hunt, male hunters often become the most supportive friends of female hunters, even if some other men are openly critical of female hunters.

This data highlights the somewhat paradoxical position in which many female hunters in rural Vermont find themselves: they are a vastly outnumbered minority in a male-dominated activity, yet through their participation in hunting they bond with men and sometimes estrange themselves from other women.

Cultural Change and Political Resistance

One of the most uniform and intriguing aspects of my research on gender is the lack of political consciousness expressed by female hunters. Generally speaking, women described their participation in hunting as something that is deeply personal and not associated with any greater concern for gender equity in the United States. Although a few women did articulate a more "political" or "feminist" conception of their hunting, more often than not the women I spoke to would, in fact, cringe at the idea of being labeled a feminist. These women also tended to downplay the negative reactions of men and emphasize the positive ones. Very often, conversations with female hunters would express two seemingly contradictory themes: the acceptance they feel in the male-dominated hunting community and their frustration with the doubt and harassment that they received from men. Almost invariably, however, the former sentiment would become the dominant representation.

One woman I came to know particularly well during my fieldwork serves as an excellent example of the complex social position that female hunters occupy and the sometimes surprising ways they reconcile the dilemmas they encounter. One afternoon, while I was visiting her at home, she began talking about how much she resented the constant leering that went on in many of the local general stores and sporting goods stores in the Addison County area. She said that when she entered these places she felt like she was expected to take off her clothes so the guys could have a better look at her body. At the sporting goods store where I worked, for example, she felt so uncomfortable that she would stay in the car while her husband went in and bought the hunting and fishing supplies she needed. I later came to learn that many women in the community felt similarly about this store.[9]

After a while, she decided to open a small grocery store of her own—one that would not be dominated by rude men. "I wanted to go to a place, and I wanted a place where it's comfortable for kids, where it's comfortable for women, and where it was comfortable for guys." And it worked. While I was living in the area her customers ran the gamut from inveterate male hunters to nonhunting mothers making a quick stop for some groceries on their way home from work. The store carried a wide variety of magazines, ranging from home decor to hunting and fishing, and the conversations were equally diverse.

Considering her motivations for opening the store, I initially found it ironic that she displayed a disparaging photograph of Hillary Clinton next to the cash register. She also bristled at my joking suggestion that she seemed like a pretty good feminist to me. After I had lived in Vermont for a while, however, I no longer found any of this ironic, because I began to realize the simplicity of my former assumptions about the relationship between political affiliations and personal desires.

Yet despite the rampant sexism this woman felt in town and the skeptical looks she received at the deer checking station, she continued to invest great value in her male friendships.[10] She once said, "I love men. I get along far better with men, like I told you the other day, than I do with women. I can't sit down and discuss how to hem pants, and what to cook tonight for dinner, washing curtains and all this garbage. You know, that's just not my style. But you wanna sit down and talk hunting and I'll sit here all night and talk hunting."

Once again, here we see a woman who, despite a large number of negative experiences with men, ultimately claims to appreciate their company more than that of women. The bonds of common practice and sentiment she shares with certain men foster the growth of relationships that are far more meaningful than those she usually has with women. I must reiterate here that the social position of female hunters in rural Vermont *is*, in fact, paradoxical. As key actors in an unfolding drama involving historic shifts in gender expectations, they navigate a social terrain that presents them with both daunting obstacles and surprisingly easy routes of passage. As a result, the seemingly contradictory stories they tell are actually accurate descriptions of their everyday experiences and feelings.

Women were not alone in reporting mixed emotions and internal conflicts. Men, too, expressed contradictory feelings about female hunters. It might have been a joke about them or a doubting glance at a checking station from a man who generally supported female hunters. On the other hand, a man who generally expressed antagonistic ideas about women who hunt might soften his position. For example, I spoke to one man who, in the course of a short conversation, went from criticizing female hunters to reflecting on how nice it might be to have a wife who hunts and could therefore share in such an important aspect of his life.

In his noteworthy book *The Meanings of Macho: Being a Man in Mexico City*, the anthropologist Matthew Gutmann points out an ideological inconsistency among his informants that is extremely similar to what I have noticed in Vermont. He ultimately utilizes Antonio Gramsci's concept of "contradictory consciousness" to great effect as a way of characterizing people who are simultaneously influenced by "traditional" gender ideologies *and* in the process of hammering out new understandings of gender in the practical activities of everyday life (see Gutmann 1996, 22, 75, 93). He argues that this tension between what might be thought of as competing ideological camps is often the impetus for social change.

The story of the hunter who opened her own store provides a convenient bridge to the next issue I want to discuss with regard to female hunters: *resistance*. As problematic as this concept can be, resistance is a social process that merits serious attention. The term is generally used to indicate social scenarios in which a subordinate group of people defies the hegemonic pressures exerted by a dominant group. Resistance studies, like practice theory (see Ortner 1984), assume that culture is an ongoing process of political struggle. As John Fiske writes in the opening lines of *Reading the Popular* (1989), "Culture (and its meanings and its pleasures) is a constant succession of social practices; it is therefore inherently political, it is centrally involved in the distribution and possible redistribution of various forms of social power" (1).

As I've suggested, resistance studies are not without their own set of problems. Most significantly, as Sherry Ortner explains (1995), they are often "ethnographically thin" and heavily dependent on a certain sense of romance associated with oppressed peoples (see Knauft 1996 and Abu-Lughod 1990). Having been initially exposed to resistance studies by way of cultural studies, I have long been wary of the tendency to too easily assume that resistance (as argued by Foucault) is present any time one locates acts of domination. I have also struggled with arriving at a working definition of resistance. What constitutes a true act of resistance? What degree of knowledge must the subordinate possess of the processes and agendas of the dominant to define certain actions as, in fact, resistant?

Specifically, are the female hunters in Vermont providing an example of resistance? I believe so. At some point during my fieldwork I became aware of the kind of tenacity it must take, as an historically marginalized minority in the hunting community, to be a female hunter in rural Vermont. It also became clear to me that, regardless of the personal politics or intentions of these women, they were changing their social world through hunting. Whether we are referring to the woman who opened her own store to create a sexually egalitarian shopping experience, the woman who did not allow herself to be intimidated after being followed into the woods by male hunters, all the women who tolerated checking station derision and the discomfort of entering local sporting goods stores, the hundreds of women who have signed up for a BOW seminar, or the woman I know in Lincoln who has won several buck pools to the chagrin of her male competitors, the common denominator is women refusing to succumb to the expectations of dominant gender ideologies as they pursue their personal desires.

Because these women clearly rejected the received wisdom regarding the appropriate conduct of females in rural Vermont, they committed acts of political and cultural resistance. The fact that most of them claimed that their hunting was a personal matter, did not define themselves as feminists, and voiced no intention or interest in "making a statement" does not reduce the importance or the resistant character of their actions as individuals or as a group. Resistant groups need not be aware of the broader consequences of their everyday actions. The only thing these women need to be conscious of is that they are swimming upstream against the male-dominated currents of American culture. Whether they thought that their hunting would lead to more female hunters in the future or more female politicians is of little import when judging whether or not their everyday practices are resisting a dominating force (contrary to Fegan 1986). As Mary Zeiss Stange writes in *Woman the Hunter*, "Whether they are self-conscious about it or not, whether they call themselves feminists or not, they are rewriting history" (1997, 186).

Male Hunters and Masculinity

There is a widespread perception in the United States that hunting, traditionally undertaken in the absence of women, is one of the supreme hypermasculine activities practiced in this country, and to many it seems that male hunters must be obsessed with expressing their masculinity through violence.[11]

Yet if it were not for my own questioning, the topic of masculinity rarely would have come up in the conversations I had with male hunters in Vermont. They did not claim that hunting should be reserved for males, that it helped them feel more masculine, or that it helped them prove they were men to themselves, their friends, their wives, or anybody else. Clearly, this does not mean that hunting is not masculine, but it does tell us something about hypermasculinity, which I define here as an aggressive, misogynistic masculinity that takes traditional attributes of masculinity to an extreme through conscious performance and actively seeks to denigrate women and exclude them from socially prestigious activities. Even though hunting is clearly a male-dominated activity in rural Vermont, and one that certainly does epitomize a certain brand of rural masculinity, I believe it is inaccurate to assume it is a hypermasculine social practice.

Ironically, while staying at a deer camp in Ripton, I witnessed a large crowd of hunters joking about what *they* considered to be a hypermasculine performance. The object of their ribbing was a particularly muscular man who was fond of wearing tight shirts. The other men assumed this was done with the intention of displaying his extremely well-conditioned body, something that is of little concern to most of the male hunters I know in Vermont[12] and thus was viewed as an extremely conscious performance of masculinity. On the other hand, in suburban America, with its near obsession with health clubs and infomercial fitness gurus, a well-muscled man in a tight shirt might not strike people as particularly hypermasculine, whereas a man with a hunting rifle on a rack in his pick-up would more likely seem to be "out to prove something" about the status of his masculinity.

The point here is that the association of men and hunting is so elementary in rural Vermont that it is often left unsaid, which is not to deny that hunting is extremely important to the construction of masculinity. To be sure, hunters generally celebrate such traditional markers of American masculinity as self-reliance, courage, and a zest for outdoor adventure; in addition to symbolically linking themselves with the exploits of frontier survivalists (Kimmel 1996; Altherr 1976). Moreover, the value invested in these rugged qualities only increases among the men who do the most hunting in the United States—men involved with manual labor (Heberlein and Thomson 1991; Heberlein 1987).

The correlation of challenging physical labor and masculine identity is a common theme in "working class" ethnographies (Halle 1984; Willis 1981; Applebaum 1981) For example, in his study of the culture of construction workers, Herbert Applebaum notes that "part of the culture of construction is the satisfaction of doing 'manly' work, winning out over the elements and showing persistence in the face of adversity" (1981, 109). In his ethnography of a chemical refinery in Linden, New Jersey, the sociologist David Halle argues that leisure activities are critical to the maintenance of what he calls a "culture of male friendships" (1984, 35). He writes: "It is hard to exaggerate the importance of various kinds of sports in the lives of these chemical workers. More than half are enthusiastic fishermen and hunters, and this is true regardless of age or marital status" (39). And in the classic ethnography *Learning to Labor*, Paul Willis observes that "difficult, uncomfortable or dangerous conditions are seen, not for themselves, but for their appropriateness to a masculine readiness and hardiness" (1981, 150).

To summarize, I would submit that hunting in rural Vermont is not a hypermasculine practice in which men engage in excessive displays of masculinity that would not be acceptable in everyday life. Although hunting clearly embodies many aspects of traditional American masculinity and seems to many urban nonhunters to be a blatant performance of aggressive masculinity, my research in rural Vermont did not expose any significant relationship between hunting, excessive celebration of masculine virtues, or, very importantly, misogynist sentiment.

Clearly, as is shown by the testimony of female hunters, some male hunters behave in a sexist manner and are opposed the idea of female hunters, and others sometimes express defensiveness and ambivalence regarding the prospect of women participating in hunting. This, of course, should not come as a surprise, since these men are involved in a dramatic shift in gender expectations and Euro-American hunting has "always" been unquestionably associated with masculinity. On the other hand, many other male hunters are not opposed to the idea of women hunting and are eager for their wives and daughters to join them in the field. Consequently, I must conclude that the notion that contemporary hunting in Vermont is generally a means for men to engage in hypermasculine activities is simply not supported by my data. The next chapter, on deer camps, will add additional support to this position.

6

Deer Camp

Incomparably the best, the most graceful, the most beautiful among the game animals familiar to us was the white-tailed deer. It followed automatically that deer hunting was the noblest form of outdoorsmanship. And deer hunting was the only form of vacation my father ever willingly took.

RONALD JAGER, *Eighty Acres: Elegy for a Family Farm* (1990)

Deer camp may be the most mysterious aspect of hunting in Vermont. Almost universally off-limits to women, deer camps are often thought of as backwoods fraternities where heavily armed men party under the pretense of hunting the cagey white-tailed deer. Drinking, violence, sexual escapades (both hetero- and homosexual), and a general appreciation for debauchery is the image that deer camp brings to mind for many non-hunters and hunters alike. While I would never attempt to suggest that there is no alcohol-induced social interaction at deer camps in Vermont, my experience certainly does not conform to the stereotypical representation that informs much public opinion of deer camp life.[1]

These camps are usually small houses or cabins located in or near hunting areas, where a group of hunters will live during the deer season. For some men this may mean staying for the complete two-week rifle season and then returning in December for another week of blackpowder season. Others may only get enough time off work to go for a long weekend. Either way, deer camp is the highlight of the year for many Vermont hunters. Just as people on the East coast will often use all of their vacation time to stay at a beach house for a few weeks during the summer, many Vermont men take their entire vacation during the rifle season for deer that begins in the middle of November and ends the weekend after Thanksgiving. Deer camp is, in many ways, one of the more reliable aspects of life, and many men I talked to shuddered at the prospect of not being able to attend. As a matter of fact, one man, whose camp I visited, told me that he had the minister presiding over his wedding read "except during deer season" into the more binding sections of the vows so he would be free to go to deer camp, without argument, for the rest of his life. (I have since been informed by University of Vermont folklorist Dick Sweterlitsch that this is a widely circulated story generally considered to be apocryphal.)

Very few hunters are occasional deer camp attendees; they either go to camp or they do not. As a result, a fairly consistent group of people end up spending time together each fall for a good portion of their adult lives. Old friends ride out the unpredictable waves of life, meeting each year at camp to "put life on hold," as one of my informants put it—if only for a weekend. And because deer camps usually run in families, the camp involves fathers, sons, and grandfathers coming together to create lasting memories and strong bonds between generations. Like rings in the trunk of a tree, each trip to camp marks the passing of another year, and, over time, deer camp becomes an important place for both the making and celebrating of a family's history.

Deer camp is a quintessential leisure activity. It provides men with a much appreciated respite from the everyday working world and a chance to relax in the outdoors with friends and family while doing something they find challenging, exciting, physically invigorating, and personally meaningful. And, as we will see, deer camp also offers people the opportunity to experience alternative modes of social interaction.

Testifying to deer camp's importance as a leisure activity, the men at camp often described a rigid distinction between everyday life and life at the camp. As one man commented during an interview two months before the season, "It's another world." This theme is reaffirmed by comments mentioned earlier concerning life "on the mountain" versus "in the flats" and the use of the term "re-entry" to refer to going back home from camp. Another man, the patriarch of a camp in Ripton, said that at camp "you can be yourself."

In sum, deer camp is much more than a reprieve from the responsibilities of work. It is a time when life is stripped down to what, for many of these men, seems to be its bare essentials: family, friends, and the natural world. As such camp becomes a rejuvenating alternative to life in the modern world that can be so devoid of continuity, lasting relationships, family history, deep experiences of home, and a connection with the processes of the natural world.

An entry from a journal in a mountain deer camp located between Lincoln and Bristol attests to these sentiments. This particular camp is a small, rustic one-room cabin with two bunks and no lock on the door. Visitors are encouraged to contribute to the journal that is kept in the bread box.

4/20/95 Bad week. At work everything seem to go bad at work. Time to get away from people, so here I am at [name of camp]. Just me and my dog [name of dog]. No phone, no people, just peace and quiet
    . . . Saw three turkeys at my deer stand this morning, the dog had a great time chasing them.
    The camp seem to have gone through another winter with flying colors. I hope this stands forever!! I really love this place, good place to regroup and think about your problems.
    Well its now Sunday morning. I guess the Chip Munks want their house back to. I guess I will go back to the rat-race.

It is widely thought, by hunters and nonhunters alike, that deer hunting is not the top priority of men at deer camp and, to some extent, I agree with this assessment. Judging from my experiences at nine different deer camps, the importance of the social interactions of the camp cannot be underestimated. Many people go to deer camp hoping to get a deer, but clearly not expecting to succeed. For them, the greatest satisfaction of deer camp is clearly located in the camaraderie it offers. It is not unusual, for example, to meet people who only kill a deer once every ten or fifteen years. On the other hand, there are certain men who take the hunting very seriously, fully intend to get deer, and would be disappointed if they did not. Men tending toward this perspective, however, are no less appreciative of the camp camaraderie. They simply place a high value on actually killing a deer, in addition to socializing with their friends and family. In my case, since eight of the nine camps were located close to the homes of their members (who, consequently, did not necessarily have to go to camp in order to hunt the same land), I am led to believe that the key factor that led men to attend camp was an appreciation of the social setting.

Moreover, though camps may be shrouded in "mountain man" romance, they are very often located in areas that do not offer prime deer hunting habitat. This is particularly true in a heavily forested state like Vermont, since deer population density actually decreases after a forest reaches about twenty years of age. For example, in 2000 the deer harvest per square mile in the mountain town of Ripton was 1.11. In nearby Lincoln, another town where I attended deer camp, the harvest was 1.88 per square mile. But in Middlebury and New Haven, two towns in the Champlain Valley "flats," the harvest per square mile was 3.62 and 4.50, respectively (VFWD 2000, 4). So why do hunters choose to hunt where there are fewer deer?

Thomas Heberlein addresses this very question in his foreword to Robert Wegner's *Legendary Deer Camps*: "Most hunters choose friends over deer. The deer are nice. They need to be there. Occasionally, a big one, like Ben's Buck, hangs on the rack. The camp is where old men tell stories, and boys grow to be men. That is where is you hunt deer. Not where whitetails run around like bunny rabbits" (2001, 6–7). I concur with his assessment. Life on the mountain is chosen over life in the flats not because the chances of killing a deer at a mountain camp are higher, but because life on the mountain offers *meanings* that the flats do not. And the special quality and significance of the deer camp experience often involve the consumption of venison. As one man told me, "Without venison you wouldn't have a deer camp." Venison becomes a sacrament that bonds man with man, man with deer, and man with Nature. It is the very embodiment of life on the mountain.

It is also important to recognize that hunters have to obtain permission to hunt on much of the agricultural land in Addison County, whereas anyone can hunt in the Green Mountain National Forest. There may be people who attend deer camp, but would rather have the chance to shoot a "rackasaurus" munching on soybean behind somebody's barn. I am inclined to believe, however, that people attend deer camp primarily because they want to go to deer camp. As I mentioned earlier, most of these men could easily hunt the same land that they hunt from camp without actually attending camp because they usually live within a half hour of their camps. Some men make the conscious choice not to join deer camps, even if they want to hunt in the mountains, because they feel the party atmosphere will reduce their chances of killing a deer. The people who go to deer camps year after year would not miss it for the world.

Many writers, primarily nonacademics, have tried to capture the special allure that brings men back to deer camp year after year (for a comprehensive bibliography on deer camps see Wegner 2001). The most substantial of these are Robert Wegner's cultural history of American deer camps, *Legendary Deer Camps* (2001), and John Miller's photojournalistic exploration of hunting in northeastern Vermont, *Deer Camp: Last Light in the Northeast Kingdom* (1992).[2] Because they approach the subject differently, these books are well worth reading as companion pieces.

Wegner provides rich descriptions of eleven notable deer camps from the Dakotas to Mississippi, exploring the camps of social luminaries such as Theodore Roosevelt, Aldo Leopold, and William Faulkner, as well as ordinary hunters like those I met in Vermont. Through engaging text accompanied by a wide array of visuals, *Legendary Deer Camps* offers a unique look into the camaraderie, lore, humor, and romance of these camps. Wegner, a hunter himself, displays a keen eye for the nuances of the hunt, providing the reader with details on gun selection and wood-craft throughout the book.

Miller's book is more journalistic in its approach, and therefore more similar to works of ethnography. Just as many anthropologists do, Miller enters a social context that is far removed from his own personal experiences. Not only is he a nonhunter, he mentions more than once that he prefers female to male friends (1992, 1, 7). *Deer Camp* should be appraised as two separate productions, one narrative and one photographic. It is probably better known for its dozens of evocative black-and-white photographs than its narrative and has assumed coffee-table status in the Northeast. The more I look at this book, however, the more I am left unsatisfied by the photography. Generally, the photos exude an ominous tone, and an eerie silence seems to descend over the book. Stone-faced hunters and gray skies suggest impending doom—presumably the doom of deer camps. Perhaps that explains the subtitle's allusion to a setting sun.

While deer camp attendance may very well decline in coming years, Miller's success at creating deeply nostalgic black-and-white images actually muddles his attempt to represent contemporary Vermont deer camps. Where is the camaraderie, the action, the passion, and the sensuality of deer camp life that Miller so deftly portrays with his prose? And where is the laughter? Anybody who has ever been to a deer camp knows that humor is one of the cornerstones of the experience. Why, then, after five years of traveling to deer camps does Miller provide only a handful of photos showing groups of socializing men with smiles on their faces? Instead, many of the men in this book look like they are at their father's funeral. This omission is all the more startling since Miller describes a number of festive evenings he enjoyed at camps during his research. He clearly knows that deer camp is a celebration of life, not a worrisome anticipation of death. His narrative communicates that, but his photos generally do not.

The most abundant sources of deer camp writing are newspapers and magazines. Topics range from the increasing presence of women at deer camps (Gwizdz 1999) to the environmental impact of commercial camps in the West (Clifford 1999). Most of the articles stick to the recurring themes of male bonding, humor, family, the maturation of boys, the importance of elders, communing with nature, and reverting to the ways of the past (for example, Hennessey 2001; McIntyre 2001; Jones 2000; Heller 2000; Robb 1999; Hill 1998; van Zwoll 1998; Quinnett 1994; Bashline 1982; Chamberlaine 1981)

Camp Social Life

> Once you walk through that door with that gun in your hand everyone's equal. It doesn't matter how much money you make or how poor you are.
>
> Deer camp owner in Lincoln

Homosocial male groups, whether athletic teams, military outfits, country clubs, or fraternities, are often thought of in both academic and non-academic circles as being sites of misogynistic, hypermasculine practices that place great emphasis on ranking, tests of masculinity, and initiations (Sanday 1990; Kimmel 1996; Lepowsky 1993). Likewise, it is not uncommon for people with no direct personal association with deer camps to assume that at least some of these characteristics will be present in those settings as well. Before my fieldwork I was no different. All I knew was that I would be driving up into the mountains and dropping in on a bunch of guys with high-powered rifles and a few cases of beer who went off to stay in the woods without women while they hunted deer. How rowdy would it actually get up at these places that are often mockingly referred to—even by fellow hunters—as "beer camps"? Drinking contests? Wrestling matches? My friends down in town wished me well.

As it turned out, the atmosphere at deer camp was much more reminiscent of a commune than a locker room. Compared to the sort of competitive male atmosphere conjured up when one thinks of the military or male athletic teams, there was a conspicuous lack of ranking and competition at the deer camps I visited. No arm wrestling, no spontaneous boxing matches, no initiation ceremonies. The closest thing I experienced to an initiation ceremony was when I was offered some very hot mustard at one camp and impressed the guys when I said I liked it and asked for more.

Killing a deer, of course, could be thought of as the ultimate initiation, since it creates a bond of highly valued common experience between successful hunters. Not surprisingly, a hunter's first deer is often a very important event in his or her life, although I heard of no special ritualistic events that accompanied a first deer.[3] For the most part, it seems that the significance of a first deer lies in the greater inclusion and adult legitimacy that the newly successful hunter feels around other hunters. As a man in

Rochester put it, "I finally had a story." He explained that he had hunted for the nearly twenty years before he killed his first deer, and one of the worst parts was never having a story to tell the guys up at deer camp. So momentous was the event when it finally occurred that he replaced the wedding picture in the frame on his dresser with a photo of that long-awaited first buck.

Work at camp was usually split up relatively equally, regardless of age or hunting ability. It was expected that everyone would pitch in on the chores that, depending on the individual camp, might include bringing in firewood, cooking, washing dishes, and fetching water. At certain camps, the older men might be exempt from this relatively egalitarian division of labor; the younger hunters often showed a very earnest and appreciative deference to some of these men who were old enough to be their grandfathers. Nevertheless, even these older guys seemed intent on doing in their fair share of camp labor.

Ultimate authority generally came down to age (if there were multiple generations at the camp) or camp ownership, but there were so few occasions where anyone tried to exercise any real authority over anyone else that I rarely witnessed any display of power, much less a serious personal confrontation. One example of a rare exertion of camp authority involved the common prohibition against bringing loaded weapons inside camp. A man arrived at camp with a loaded handgun, and even though he was a good friend of the camp's owner he was ultimately asked to leave because he refused to unload the gun.

In terms of the general social dynamics of ranking, authority, and the division of labor, the deer camps that I visited were certainly not another case of a cultural practice reproducing the capitalist social structure. In addition, as I mentioned earlier, the competitiveness that is so often revered in the United States as both the driving force of capitalism and a highly desirable personality trait was noticeably absent as well. This fact was not lost on the men at camp. Indeed, they actually took precautions at several camps to avoid excessive competition. The most commonly mentioned examples were keeping women out of the picture (because the presence of women, I was told, often led to competition and fighting among men) and limiting the amount of money, if any, that could be wagered on the card games that are often played at deer camps.

When considering the social structure of deer camps it is important to keep in mind that, for the most part, the men at these camps are not white-collar workers. Like most American hunters, they were generally middle-class, high-school graduates who worked in the trades or agriculture. As social scientists have repeatedly shown (Eckert 1989; Willis 1981; Foley 1990; Applebaum 1981), working-class people are often decidedly more oriented toward egalitarian social relations than highly educated, upper-middle-class people who are well acquainted and quite comfortable with "a materialistic culture that is intensely competitive, individualistic, and unegalitarian" (Foley 1990, xv). In part, this is due to the lack of socioeconomic and educational differentiation among members of the working class. In the case of skilled craftsmen and farmers, egalitarian social structures are closely related to the nature of their work, since there is a great deal of independence associated with their work and they are generally not concerned with things like the corporate ladder or upward mobility. As the sociologist H. J. Morgan writes in *Discovering Men*, "The mobile career orientation relates to themes of striving and competitiveness and individuality; the relatively immobile, lower-class occupational statuses may play a part in the construction of masculinities based on group solidarity, fraternity and loyalty" (1992, 84). Consequently, I would contend that the solidarity and equity of deer camp is both a reflection of traditional working-class culture and a reaction against the stratification of American society.

This is not to say that the white-collar workers at deer camps adhere to a strict social hierarchy. One camp I attended, in fact, was comprised almost completely of white-collar workers, and it was very similar in atmosphere to the other camps, notwithstanding the cell phones to check stock prices and dips in the hot tub on the deck. One could argue that this particular camp is actually a greater departure from everyday life than the more working-class camps because its social relations are further removed from what these men experience in their everyday lives.

Camp Humor

A fundamental characteristic of most deer camps is humor. Particularly at camps that house more than two or three men, the generation of laughter seems to be a major priority. Often the topics seems to be chosen at random, but one persistent theme of the joking and horsing around centers on human ineptitude. Hunters were very often portrayed as clumsy buffoons stumbling around the woods after deer who had long since vanished, or even defecating in their pants as they encountered a more dangerous animal like a black bear. I was the butt of some of this joking at a camp just south of Canada, in the extremely rural area of Vermont called the Northeast Kingdom.[4] As we were sitting around the dinner table one night, one man asked me if I had ever heard of a kind of wild vegetation called *milermore*. Taking him quite seriously, I said I didn't think I had. He went on to say that it grows pretty thickly in the area around the camp. Once again, still completely unaware that I was being put on, I admitted that I didn't think I knew what milermore was. Not wanting to come off like a complete greenhorn, however, I was quick to extend my neck further into the snare by adding that we did have tons of multiflora roses in the area of New Jersey where I grew up. Finally, out came the punch line. "Yeah," he said, "we've got it all over the place up here. When you step on it you can hear it for a *mile or more!*" Anybody who has ever attempted to get very close to wild animals, for any reason, can probably relate to the humor in this joke, not to mention the humor of my own gullibility.

As surprising as this kind of "inept woodsman" humor may seem, one might make the argument that this kind of humor should actually be expected, since it deals with the critical issue of self-sufficiency. As I have noted, an appreciation for self-sufficiency is extremely pervasive among all the hunters I have met in Vermont—male or female. So the idea of self-sufficiency constitutes an important part not only of Vermont hunting culture in general, but also of conceptions of masculinity and femininity as well.

Another theme that was noticeable in camp joking was sexuality. Similar to the "inept woodsman" genre, these jokes often poke fun at human sexual misadventures. For instance, one man—in about a five-minute monologue that focused on impotence—joked that his penis was "ingrown" and "needs a pump" to maintain an erection, and that, as a result of his problems, he has now resorted to "putting a string on it."

Homosexuality was also a subject of joking banter at some camps. These jokes were not made in an aggressively homophobic manner, though some may argue that homophobia is at the root of these comments. Generally, this involved men joking about the homosexual tendencies of another guy in camp or joking that they were sexually attracted to another member of camp. For example, a note left at one camp read, "Good thing [man's name] isn't here yet. He's been away from [his wife] for a week. Zip your sleeping bag in case he shows up." Two different cartoons displayed on the wall of one camp depict a buck raping a hapless hunter. Under one of them it reads (as I recall it) "That's what you get for killing all the does"—an obvious reference to the ongoing debate about the wisdom of hunting for female deer. The other caption reads "Here's your deer meat!" And a couple of years ago I heard somebody make a passing comment about the "queer union" bill in Vermont. Since I neglected to follow up on that comment, I cannot be sure of its intended meaning or the feelings on civil unions held by the person who said it.

Nevertheless, I never witnessed the kind of extreme homophobia described by Peggy Sanday in *Fraternity Gang Rape*—which, she argues, actually leads to acts of sexual violence against women as a means of "proving" one's heterosexual manhood. I slept in a double bed with another man at one camp, and this sleeping arrangement was suggested without the slightest hesitation. Judging from the bedding options—mostly double beds—it must have been a fairly common occurrence. I found the situation somewhat ironic since I had not slept in bed with another male for over ten years and I had always considered myself fairly relaxed about these kind of situations.

Looking back, I simply had never heard so many men openly admit and joke about their own faults and weaknesses as I did during my stays at deer camps. This often self-deprecating humor, of course, fits the relaxed, friendly, and noncompetitive spirit of camp life. The atmosphere was so noteworthy to me that in my fieldnotes for one November day in 1997 I wrote, "Everybody doing their part. No bickering. Nobody's defensive. No competition." I remember quite clearly, for example, seeing one of the state's great hunters sitting in his longjohns with tears in his eyes as his fifty-year-old nephew told me how scared they all were about his uncle's recent heart surgery. Indeed, I must emphasize that for many of these men at deer camp it is difficult to overestimate the importance of the relationships they maintain with their male friends. As one man put it when talking about people who do not have a cohort of male friends, " I don't know how they survive." He also said he felt sorry for his wife because she does not have an equivalent group of female friends.

Perhaps surprisingly, it seems that the displays of emotion and admissions of weakness that occurred at the camps were directly related to the *absence of women*. In the course of interviewing a man whose camp I would later visit, I asked him why he found camp so enjoyable; the first thing he said was "You can be yourself." I wondered if "be yourself" was just a euphemism for "walk around drunk in my underwear," but from my first day at camp it was clear that this was not the case. It was if the burden of self-assured, all-knowing, fearless, and dominant American masculinity had been lifted from their backs at the door of the camp. In the company of trusted friends, these men were able to speak freely of fear, insecurity, and vulnerability in ways that may not be possible during their everyday lives.

He was not the only man to tell me this. Others said it as well, while implicating the presence of women in more direct ways. Among the men I dealt with, it appears that women are viewed—fairly or unfairly—as part of an everyday world that expects them to live up to traditional standards of masculinity, and part of the affection many men have for camp is that it provides a culturally sanctioned way of escaping these pressures. But that's not the whole story. Clearly, deer camp is not simply as escape from the complicated burdens of masculinity. It also satisfies a desire to share sensitive, egalitarian relationships with other men; precisely the kinds of relationships that can be so difficult to develop in place as homophobic and competitive as America is today.

Misogyny?

An obvious question to ask about deer camps is whether or not they constitute a misogynistic environment. Are they, in fact, rural America's version of the "men's houses" we so often read about in the anthropological literature (Lepowsky 1993; Godelier 1986; Herdt 1981; Murphy and Murphy 1974)? To be sure, this is an extremely complicated issue. From what I have observed, however, I would contend that antifemale sentiment is not one of the fundamental characteristics of deer camps in Vermont. Women are not regularly ridiculed, nor are men often ascribed female traits as a form of derision—the standard operating procedure of many sports coaches, for example. On the contrary, it is not uncommon for men at camp to speak fondly of their wives and, as we've seen, joking at camp often revolves around the ineptitude of men.

The aspect of deer camp behavior that stands in strongest opposition to the conclusion that camp is not a misogynistic environment is the common presence of pornography at camp—usually in the form of popular magazines such as *Playboy* and *Penthouse*, which were usually in plain sight unless women or children were present. Not all the men looked at these magazines in my presence, but many did. I was particularly interested in the comments that men made about the women pictured in the magazines. Generally, the few comments that were made (while clearly objectifying the women) were not abusive and degrading, but dealt with how beautiful or alluring the women were. This may seem like an obscure and inconsequential observation, but there can be great variation in the ways that men describe women who are sexually objectified—whether in pornographic magazines, working at strip clubs, or anchoring television newscasts.

One appropriate example, since it also involves an all-male environment, comes from a prestigious New Jersey country club where I bartended immediately after completing my initial fieldwork in 1998. As is the case with the legendary men's club, the Bohemian Grove (Kimmel 1996), many of this club's members are prominent men in American business, politics, medicine, and law. In true country-club fashion, the television stations of choice at the bar were any that happened to be broadcasting golf, or CNBC—a popular financial sector channel that closely tracks events on Wall Street. Several of the reporters on CNBC are women who meet stereotypical white, patriarchal, American standards of feminine beauty. When these women were giving their reports on the latest advances in tech stocks or the continuing troubles with the Japanese markets, it was not uncommon for men at the bar to casually remark, "She swallows." These comments, of course, fall into that great tradition of men morally condemning women for fulfilling their own sexual fantasies. I always wondered if these men would be saying such things if these women were not in positions of expertise, if these members of a blatantly sexist and racially prejudiced club were somehow threatened by an attractive young woman giving them financial advice. Compared to the routine behavior of these wealthy country-clubbers, the men at deer camps were relatively mild, seemingly centered more on admiration and fantasy than denigration.

Another factor that must be considered is the way these men treated women in everyday life. As I have noted, the deer camps I visited were not generally places that were oriented around the debasement of women. Furthermore, I was often impressed with the quality of the relationships the men had with their wives and families. All I can responsibly do at this point is make judgments based on the interactions I experienced. With that as my guide, it does not seem like the existence of pornography was indicative of a general antipathy toward women.

In many respects, then, my deer camp data offers a vision of masculine sociability that is in opposition to what has been provided by some well-known studies of all-male groups in Western settings (Kimmell 1996; Herzfeld 1985; de Almeida 1996; Sanday 1990). For example, in *The Poetics of Manhood* (1985) Michael Herzfeld describes a rural Cretan male sociability that is deeply concerned with competition, the evasion of female pollution, and a rigid distinction and adversarial tension between men and women. Sexual power and insecurity are such key organizing themes that café card games become metaphors of male sexual domination of women. Likewise, in *The Hegemonic Male* (1996), Miguel de Alameida documents similar practices as the café becomes a critical arena for Portuguese men to enact grand cultural performances of aggressive and dominating masculinity. Peggy Sanday's chilling account of gang rape among college students, *Fraternity Gang Rape* (1990), provides another example of the violent denigration of females by homosocial male groups. I have never encountered any of the adversarial and violent tendencies that these books so effectively chronicle. Deer camps can certainly be described as separatist, but to characterize them as misogynistic would run directly counter to my current data.

My research experiences among deer hunters in Vermont more closely resembles those Mathew Guttman describes in *The Meaning of Macho: Being a Man in Mexico City* (1996), since what I encountered was not what the academic or popular literature on masculinity led me to expect. Guttman writes: "Later, as I reviewed the social science literature on Mexican men and masculinity, the topic of my research became clear: widely accepted generalizations about male gender identities in Mexico often seemed egregious stereotypes about machismo, the supposed cultural trait of Mexican men that is at once so famous and yet so thoroughly unknown. Even when I read of individuals and groups who for some reason did not fit a pattern of machismo—which, however it is defined in the social sciences, generally carries pejorative connotations—those cases were routinely judged to be unusual" (1996, 12).

It has often been stated in studies of masculinity that emotionally sensitive relationships between men are among the greatest casualties of American manhood (Kimmell 1996). I will conclude, then, by briefly describing an evening that would prove to be one of the turning points in my understanding of the great value many members of deer camps place on their membership in a community of intimate male friendships. It was during the first week of deer season, at a camp I was staying at in Ripton, and it was the night of the annual party that this camp throws. About thirty men (some with their sons) came up to the camp to eat dinner and socialize. The atmosphere was positively festive—plenty of venison, beer, and laughter. I had a great time, from both a social and an anthropological perspective. I gained a reputation for having an insatiable appetite as I polished off more than my fair share of potatoes and venison; much to the chef's delight. And I had a great conversation with a guy who looked like Tom Cruise with calluses about his wife's disappointment over their teenage daughter's decision to be her dad's new hunting partner.

But the most striking parts of the evening came when two members of the camp, at different times, made a point of telling me how much it meant to them to see all the guys having a good time together. One of them mentioned the lack of "bickering" between the men, and the other guy—the 220-pound carpenter who had done most of the cooking—commented on the great satisfaction he gets from simply watching everybody eat. I think what made these moments so memorable for me was that both of these men were sitting quietly by themselves, just admiring the scene that was unfolding around them. They seemed to be taking in all that they could get.

7

Illegitimate Killers

Watershed moments don't happen every day. That's why February 25, 2005, is an important date if you're interested in the future of hunting in the state of Vermont. What happened then was seriously odd by rural Vermont standards: an organized hunting protest. More specifically, a grassroots protest, organized by a recently minted activist group calling itself Vermonters for Safe Hunting and Wildlife Diversity, of a high-profile coyote-hunting tournament. That initial protest led to another protest of the same tournament in 2006. More importantly, it has catalyzed ongoing public debate about the ethics and meaning of hunting. The following excerpts from letters to the editor in area newspapers, along with an editorial from the *Burlington Free Press*, provide a cross-section of the kinds of topics that are being folded into the coyote tournament controversy in Addison County.

> Out-of-staters are moving to Vermont to retire, to vacation and to pursue a way of life not possible in other parts of the country. These new residents tend to be well-educated, politically savvy and environmentally aware. To many of these people, including myself, a Vermont resident for only five years, hunting is an anachronism, and not part of modern life. These "new" Vermonters will slowly begin to reshape the ethical, political and fiscal landscape of our community with their beliefs, voting practices and dollars. . . .
> My business is environmentally friendly and does not support discrimination or violence against any gender, race, religion, physical disability, sexual orientation, or in the case of coyotes, species. . . . I have made a personal commitment to boycott all businesses which support coyote hunting and sponsor killing contests. I contact other business owners, neighbors, friends, and family members urging them to do the same when I learn of a business that supports these inhumane practices. Already local businesses have lost a number of customers and will continue to lose more as a grass-roots effort to stop the carnage by boycotting offending businesses grows in strength. If people cannot be taught the value of all animals in our natural world, perhaps they can be swayed by the almighty dollar.

As a mother, it is my responsibility to raise my son to become a productive, peaceable, member of our community. I feel it is vitally important to do so in an environment that neither fosters violence, nor promotes killing as sport or leisure. As adults, it is all of our responsibility to teach our community's children compassion and a reverence for life. . . . I will not allow my child to be traumatized by piles of bloodied dead coyote carcasses laying by the roadside on "derby" days. This is the stuff of lasting nightmares. . . . If towns can pass ordinances forbidding junk cars from littering scenic by-ways [so] as not to offend tourists, I can't imagine that state tourism numbers would be boosted by visitors trying to understand why mangled animals are lining our roadsides and are dumped on the picturesque front porches of quaint general stores.

PALEY ANDERSON, Orwell
Addison Eagle, February 18, 2006

I am going to try hard to give my viewpoints to Ms. Anderson's article. This may be difficult as she has come to our State of Vermont very informed, highly educated with great wealth and political saviness, I guess with my lack of education and money I should just roll over and get out of the way of progress.

She would like to entertain us with all of her knowledge of wildlife, when in fact she and her likes are totally ignorant [of] wildlife and their habitat. They come here and build houses in some of the best wildlife habitat and force coyotes and others into our backyards, which throws the ecosystem out of balance. Yet we hunters, loggers, lumber mills and foresters are to blame for destroying the environment.

As far as the so-called killing contests go, I have never participated in one, but do not confuse me with someone who does not support my fellow hunters in trying to do something to help increase population of deer or small game as I feel our Fish and Game department is sitting on their hands and not doing what should be done to help create habitat. . . .

Mrs. Anderson, you can protect your son from us by keeping blinders on him from the truth, but if exposing him to the carnage that you say is involved in these hunts causes nightmares it is because of the lifestyle that you will only allow him to see. My family has been exposed to such "carnage" for generations and I can guarantee you that there have been no nightmares. . . .

Addison Eagle, February 25, 2006

I recently read Paley Anderson's letter to the *Addison Eagle*. Understanding that we have some differences of opinion on subjects such as hunting and fishing in Vermont I would ask you to consider the following.

I agree with you that many people come to Vermont to find a way of life that they cannot find elsewhere. . . . We don't attack people that have different points of view than ours as less educated or for being of a lesser financial stature [than] another neighbor.

Your boycott of local business is also upsetting from my point of view. Many Vermonters with points of view that differ from yours make a living in small mom and pop businesses. They are as much entitled to a means of supporting their families as you are. . . .

Mrs. Anderson, in my neighborhood when a new neighbor moves to the area we welcome them with open arms. Many times holding social events in our homes to welcome them. Is what you are advocating what we have to look forward to when someone from out-of-state moves to our neighborhoods? I certainly hope not.

Addison Eagle, February 25, 2006

To Ms. Paley Anderson, against coyote hunting: Your sentiment is fueled by emotion and lack of cultural tolerance. There is no reason or science behind your argument. You are a socially discriminating elitist with your shallow view of New Vermont. Imposing the way you feel with respect to our heritage admits your ignorance. . . .

What hunting offers can build character and responsibility in our youth. And offers a wonderful opportunity for parents to connect and talk to kids. It teaches patience, focus, safety, integrity, community and that rewards are not instantaneous upon effort. It keeps a lot of youth from serious trouble. The benefits are endless. . . .

The long term effects of your half-baked vision of a new Vermont would have horrible ramifications for game and nongame species and their habitat. But I suppose, by your article, that social discrimination is what you stand for more than active conservation benefiting the natural world.

Addison Eagle, February 25, 2006

I choose to live in Orwell for the same reason the "hunters" do—that being a love for the free country, the wilderness, the openness and a fast disappearing way of life. I would like that country back. Up until the last few years it was a peaceful place here. But the new phenomenon of hunting coyotes with uncontrolled radio collared hound dogs puts usually predictable coyotes out of their hiding places and into our backyards. Short wave radios make it possible for many hunters to show up quickly making me feel we must put our animals in and stay inside ourselves. The tournaments only make this problem worse. . . .

. . . My point is that we need to draw the line somewhere and balance the rights of hunters and landowners alike.

. . . So given the fact that the promoters of these hunts carry no liability insurance, that these hunts are anything but "traditional" and that this is an accident waiting to happen, let's try to negate the polarizing effects these hunts have on our community and reach some kind of compromise. I urge support for [House bill] 745, which simply ends these carnival-style killings. I think many people feel like I do and would like to see this controversy end.

Addison Independent, February 20, 2006

On other notes, the Coyote thing is festering again, this time fanned the by same environmentalist element that hounded the hunters last year. . . . The assault that these types are making on our hunting heritage is preposterous. The state has defined legal hunting activities in the manual and the lawbook that we all have available. Those statutes (laws) define the legal activities of the state. Ethics, lifestyle choices, personal preferences, and the myriad other things that affect hunting, such as the type, color, size, and manufacturer of your clothing, whether you hunt alone or with others, and what you do with your game after you have harvested it are choices better left to the individual than something that we should be legislating. Yet these kooks and all their envirowhacko friends now want to tell you what you have to do with your game, after you shoot it. . . . Talk about "BIG BROTHER"—I never heard of anything so ridiculous in my life. Pretty soon they will be in your kitchen of living room telling you which vegetable you have to prepare or what kind of sofa you have to have. . . . Before you know it—your rights to hunt and fish will be gone. I am not being an alarmist. The opponents HATE your sports and your right to enjoy them and, and will not stop until the world conforms to THEIR view of it. You and your guns are NOT in their world. . . .

Sam's Good News, February 22, 2006

Our own Fish and Wildlife Commissioner will not stand up and tell us that coyotes kill deer. They have sold our deer herd for the last 30 years for $10 a head. Anybody in their right mind knows coyotes kill deer; we shouldn't need to have a biologist tell us. My late father often said biologists were educated "gomps."

. . . It was reported that a cow in Castleton was killed by a coyote. Can you imagine what would happen to a child waiting at a rural bus stop or playing in his yard if he was attacked by one of these animals. Have we got to wait until some tragedy like this happens before something is done about coyotes? By that time we will be overpopulated with them and it will be too late."

Some protestors don't want coyotes shot because you don't eat them. Well, how do you want them cooked—roasted, boiled, or stewed, and then will it be OK to hunt them.

. . . There is no difference in coyote pools, deer pools, or fishing pools. All killing for sport and prizes.

I am writing this letter on behalf of my brother, [who] is having a coyote shoot and I think the hunters should give him some support.

Burlington Free Press, March 13, 2006

I am writing to express my utter disgust at the news of this weekend's coyote hunt in Addison County. This is about the love of killing and nothing else. All you so-called "hunters" should be deeply ashamed of yourselves. Any hunter who has not taken the time to educate themselves about their prey should lose their license to hunt.

Coyotes actually increase their numbers in response to threats to their breeding. So what is your excuse now? More to kill next year? More thrills as you blast away the life of a being with the same rights to existence as your own? Telling yourselves you are helping to increase the deer herd? Foolish! You are creating more, not less, coyotes.

Fish and Game, wake up! It is time to stop this outrageous practice by people who do not deserve to hunt. I am ashamed to live in a state that allows this barbaric, unnecessary and dangerous practice to continue. I will spend the weekend praying for the souls of coyote and hunter alike.

Addison Independent, March 9, 2006

What is particularly worrisome is the potential for these derbies to harm the state's traditional hunting image. For example, too many of the coyotes killed as part of derbies are discarded rather than being properly harvested for their fur. Sometimes prizes are awarded to the hunter who kills the smallest animal. That creates a negative image of the sport.

This Fish and Wildlife Board should carefully review the science and carefully construct a ban on coyote-hunting tournaments that doesn't threaten traditional coyote hunting.

After all, hunting is an important part of Vermont's heritage, as well as its economy.

Editorial, *Burlington Free Press*, March 13, 2006

From a purely academic perspective, I must admit that I wasn't exactly upset to hear about the controversy over coyote tournaments. This was the kind of controversy that my graduate-school advisors were always hoping I would report: a story with bitter adversaries, winners and losers, and maybe even some compromising information on the almighty STATE. And they always looked so skeptical when I told them that the story didn't exist. Well, now it did, and I eagerly awaited the reaction of the local community to the pressure of these protests.

The exertion of pressure on systems, whether social or mechanical, can often lead to valuable insights.[1] Tom Heberlein once explained this phenomenon with a NASCAR metaphor: What do NASCAR mechanics do to test a new engine? Do they let it idle in the garage for a little while? No, they red-line it, run it until it's about to blow up, and then they take it apart and see how the different parts held up to the demands of extreme pressure. Under pressure, then, we can often learn things about the workings of a system that may not be apparent under normal circumstances. That's exactly why public controversies can be so interesting and valuable to study. (For a fine example of this approach see Dizard 1999, which analyzes a hunting controversy in western Massachusetts.) These tournaments have forced Addison County hunters to define "hunting" at an unprecedented level of detail and now—in what must be a nightmare scenario for pro-hunting advocates—hunters are arguing with one another about what constitutes an ethical hunting experience. More broadly, these tournaments have become a site of local resistance against the invasion, control, and definition of rural space by protestors and "antis" who are often assumed to be immigrants to Vermont from more urban states. Furthermore, considering both the subordinate position occupied by "rural America" in our national cultural hierarchy and the changing demographics of rural America, this controversy demonstrates the important role that human–animal interactions play in reproducing rural Euro-American cultural identity (Fitchen 1991; Ching and Creed 1997).

A Report from Tournament Headquarters

I attended my first coyote tournament in the spring of 2007. Gaining entry was a bit tough because the organizer was committed to avoiding the controversy that had befallen the same tournament (though with a different organizer) in 2005 and 2006. On the tournament entry he had written: "This is my first year as organizer. I have hunted in all the other hunts and I intend to hunt in this one. I fully intend on keeping this hunt as low key as possible. I mean absolutely no reporters or media of any kind. I have 12 private acres well away from main road, and only want hunters or people who support hunting to be there. Good luck and remember, we are watched by people who would love to take away our hunting rights!"

After a few long phone conversations, during which I gave him my word that I wasn't, as he put it, "out to start any trouble" by writing a nasty newspaper article or exposing the date and location of this relatively underground event, he finally decided that I could come along and see what a coyote tournament was all about. As with so many other cases, it was a great help that he knew my brother, the manager at a local sporting goods store. I seriously doubt I would have been permitted to attend the tournament without that important local connection.

Tournament participants pay an entry fee and are asked to obey certain tournament regulations pertaining to the legal dates and locations for killing coyotes that will be entered in the tournament. They also learn where they should bring any "dogs" (the term tournament hunters often use when referring to coyotes) they kill for weighing. These tournaments—which have been held in Vermont since at least the early 1990s—can be quite large, sometimes with more than five hundred participants.

A hundred and thirty two people paid the $20 fee to participate in the 2007 tournament. The hunters competed for $1,400 in prize money, and participated in a raffle for a $630 rifle that the organizer purchased from a local sporting goods store with money taken from the entry fees. Eleven coyotes were killed at this tournament.

In 2008, in an effort to spread the wealth, the organizer eliminated the big-ticket raffle item (the hunting rifle) and replaced it with a large number of $25 gift certificates from local businesses. He also gave a $20 prize to anybody who killed a coyote, regardless of its size. The 2008 tournament drew 110 entrants, and approximately thirty coyotes were killed. One hunter observed that the larger number of coyotes killed probably reflected the lure of the $20 prize, which may have motivated people to drive over and check in their coyotes even if the it was too small to be "in the money."

I did not participate as a hunter in either the 2007 or the 2008 tournament. Instead, I based my research at the tournament organizer's house—which is the weigh-in station and general base of operations for the tournament—for a significant portion of a Friday, Saturday, and Sunday for both tournaments. This proved to be a wonderfully productive place to conduct research. In addition to spending important time with the organizer, his wife, and their college-aged daughter, I met and observed dozens of hunters (almost all of them men ranging from their late twenties to fifties) as they stopped by the house, either to weigh in coyotes or to attend the awards ceremony on Sunday afternoon.

Probably the single most important human–nature theme that emerged from my research at the tournaments is the extent to which coyotes are approaching inclusion in the unenviable cultural category of vermin—a good-for-nothing animal that doesn't even merit the time and effort to hunt. While it was not uncommon to hear people describe specific coyotes with words like "beautiful," or to hear hunters talk about how much respect, and even admiration, they had for coyotes, there was no comparison between the ways that people at the tournament were talking about, and treating, dead coyotes, and the way that hunters generally act around dead deer. Most people, for example, had no interest in the bodies of the dead coyotes. Only two people out of the thirty or so who showed up at the final awards ceremony wanted to take any of them home when the ceremony was finished. I was quite surprised by this, since the failure to utilize the bodies of dead coyotes is probably the most important bit of information that antitournament protestors use to support their claim that the tournaments are (as one opponent put it to me) "senseless slaughters." I thought the hunters would at least feign an interest in utilizing the coyotes, but most people didn't even want to touch them. I finally stepped forward at the 2007 tournament, when nobody else would, to help the organizer stuff the frozen coyotes in plastic garbage bags for the man who was taking most of them home. The same lack of interest was evident at the 2008 tournament. The organizer even went so far as to implore the hunters not to throw the bodies on the side of the road as they left his house, repeatedly reminding them that they would be damaging the image of coyote tournaments—and, by association, all hunters—in the process.

As is often the case around game weigh-in stations, an almost carnival-like fascination hung in the air on the Sundays of the award ceremonies, as the men and a handful of children—mostly boys—milled around the dead coyotes. (After all, part of the allure of all hunting seems to be the opportunity to be so close to wild animals, even if they are dead.) The hunters I have talked to on these occasions have been quite friendly and sometimes downright jovial, which isn't surprising because the awards ceremony are a special occasion, like a ball game or a party. A couple of people were drinking beer (which upset the organizer quite a bit), one with his jacket wide open (which I found quite impressive, since I was struggling just to keep my teeth from chattering).

I asked one man, probably in his late twenties, why we don't just try to exterminate the coyotes once and for all. After all, they kill deer, pets, livestock, and maybe even people. He responded instantly, saying he wouldn't want to exterminate them because then he wouldn't be able to hunt them. His answer resonated with all my experiences and conversations with serious coyote hunters. They enjoy hunting coyotes because their intelligence makes them challenging prey, and the fact that they aren't eaten is justified by their supposed overpopulation and the damage they are said to inflict on the deer population. Of course, this public-service rationale is important to many "varmint hunters," not just tournament coyote hunters in Vermont.

It all felt very much like a fishing tournament to me. Between the entry fees, the weigh-ins, the awards ceremony, and the coyotes being hung by their Achilles tendons from a fishing scale, the similarity to a fishing tournament was fairly easy to notice.[2] Of course, you can't throw a coyote back after you shoot it.

After the 2007 tournament, I lingered at the organizer's house for at least a couple of hours, sitting around drinking tea, making small talk, and watching TV. As I had come to expect after the long phone conversations and a few days hanging around his house, he was nothing if not a warm, caring family man. I couldn't help noticing the bond he shared with his daughter as he hugged her good-bye before she made the drive back to her college. Again and again, he urged her to be careful on the snowy roads. Having two young daughters of my own, I was particularly moved by his nostalgic recollection of the time in his daughter's life when he realized that she would soon be too big for him to pick up and hold.

As is typical of almost all of the hunters I have talked to in Vermont, he had a real connection to, and concern for, the natural world. For example, that night, with misty eyes, he told a story from his childhood about a honey-hunting uncle who cut down a beautiful elm tree on his family's property to access a cherished beehive. Either this coyote hunter was try-ing hard to prove to the visiting anthropologist that coyote-tournament organizers can, indeed, have a heart, or he was still lamenting that unfor-tunate incident. I had no reason to doubt his sincerity. He also told me about a story he had seen on a television news program about a man who had cut down a huge tree near his property line in a dispute with his town. Once again, the coyote-hunt organizer was clearly disgusted by what he thought was a senseless killing of a tree.

A little later in the evening, we were talking about fishing and I men-tioned (OK, *bragged*) to him that I caught the biggest largemouth bass in the history of a local Addison County fishing tournament back in the 1990s. He immediately asked if I had released the fish. When I told him I had, he responded with a mock-serious "OK, you can stay then," as if he would have kicked me out of his house for killing such a grand old fish.

Some might find these incidents paradoxical, or even hypocritical. Why would a coyote-hunt organizer care about a fish that he will never eat, or even see? And why would he get choked up about a tree from his distant childhood? Once again, as I have done at other points in this book, I would say that paradox depends on unmet expectations, and there really isn't any good reason to expect a coyote-hunt organizer to be a belligerent person, either toward the natural world or toward other people. This man's feel-ings about coyotes, trees, and largemouth bass—feelings that may be hard to reconcile for some—are simply emblematic of the complexity of hu-man–animal relationships.

Perspectives

There are three main players in the coyote tournament drama: supporters, opponents, and the Vermont Department of Fish and Wildlife. Arguments in support of the tournaments, as we have seen so far in this chapter (as well as in chapter 2), generally hinges on three points: 1) the threat coyotes pose to the deer population; 2) the potential threat coyotes pose to people; and 3) a person's right to hunt. Reasons for opposing the tournaments, however, are a bit more varied because there is presently a broader spectrum of people speaking out against the tournaments. Whereas very few, if any, nonhunters are strong supporters of the tournaments, opponents come from the ranks of antihunters, nonhunters, and avid lifelong hunters. In the process, taken-for-granted social dichotomies in Addison County (i.e., rural/urban, hunter/antihunter) are becoming increasingly unstable. The most common critiques of the hunts are that they are immoral/oppose traditional hunting values (because they are unnecessary, because the coyotes are not utilized, and because of the prizes that are awarded at the tournaments); that they damage the image of hunting; and that they are dangerous.

In October of 2007 I met with Jim Hoverman, the president of Vermonters for Safe Hunting and Wildlife Diversity (the group that has led the protests of the tournaments), at his home on the outskirts of Middlebury. Jim is a former college football player and was inducted into the New England High School Hall of Fame as a wrestling coach; he is also a vegan. His opposition of coyote tournaments is focused on the immorality of the hunts (he was the one who referred to them as "senseless slaughters"). As is common among animal protectionists, during both of our conversations (we also spoke by phone in January of 2007) he argued for the rights of *individual* animals, as opposed to the usual wildlife management emphasis on *populations* of animals. He explained to me that "life is life," regardless of the specific species of animals we are considering. We reached what seemed to me to be his bottom-line argument when, in reference to the coyotes killed at tournaments, he asked, "Do the lives themselves have meaning?" After that query had a chance to settle he asked another morally loaded question: "Is that what we want to teach our youngsters?"

Hoverman does not oppose hunting in general. He claims to have no problem with people who hunt for meat and freely admits that many hunters know more than he does about the habits of wild animals. Moreover, in 1994, when he gave up eating meat, he was actually raising and slaughtering much of his own food on a farm in New Hampshire. As he tells it, one day he took a walk in the woods and heard in the distance the repeated gun shots of a neighbor slaughtering some pigs and decided then and there, "I don't want to do this anymore."

Hoverman makes no claim to having a monopoly on moral living. And he acknowledges the considerable moral grayness that surrounds many human–animal interactions. But he feels that the coyote tournaments have "gone over the edge" of acceptable moral behavior because they encourage people to kill animals "to make money and win stuff." Indeed, it was his complete moral revulsion with the idea of awarding prizes to people who killed coyotes (which he first learned about through an advertisement for the Howlin' Hills Coyote Hunt that he saw at the East Middlebury General Store) that initially turned him against the tournaments and eventually led to the 2005 protest in Whiting.

Jim Hoverman's moral critique of coyote tournaments, with its explicit concern for the meaning of individual coyotes' lives, is almost always voiced by nonhunting and antihunting opponents of the tournaments. As for hunters who don't approve of the tournaments, their moral critiques are usually less intense, and they may not necessarily be opposed to killing coyotes (in a non-tournament context) to "save a deer" or obtain a nice pelt for the wall. As one hunter told me in January 2007, with all the coyotes she sees these days she's "kinda torn" about the whole issue. While she's not opposed to killing coyotes in general, she is opposed to killing them in tournaments—what she referred to as "bloodsport." She told me a story about a group of people who killed a coyote by chasing it down and running it over with their snowmobiles. Even over the phone she blew me back with her raw anger: "I went nuts! I just went absolutely nuts. That is the most hideous thing I have ever heard. Where do we get off thinking we're the most almighty out there?"

Most of the hunters I know in Vermont seem to fall into the wide area between outright opposition to and enthusiastic support for the tournaments. While I have yet to meet a hunter who would actually attend a protest of a tournament, there are those (like the woman I just described) who oppose them. Some hunters don't oppose them, but have no desire to participate themselves. Others neither hunt coyotes nor support the tournaments, but still wouldn't take the step of actually publicly criticizing the tournaments. And there are those who are primarily concerned with the bad image that these tournaments seem to be giving hunting and fear that somehow the tournaments will catalyze public opposition to hunting in general. Indeed, in the weeks before the tournament I attended in the spring of 2007, several local hunters who had participated in past tournaments were not happy with the tournament organizer's decision to hold another one. Considering all the recent controversy over the tournaments, many folks thought he was "pushing it" or "asking for trouble."

The position articulated by Vermont Department of Fish and Wildlife personnel is in neither complete agreement nor disagreement with either the pro- or anti-tournament factions. As the agency charged with providing the scientific guidance for wildlife management in the state of Vermont, the department attempts to appraise the tournaments in a dispassionate manner, judging them by their effect on animal populations rather than their moral legitimacy.

I've had numerous conversations about coyote tournaments with several extremely generous members of the Department of Fish and Wildlife, and their position on the issue is quite clear: The department does not oppose the hunts, because they have concluded that they do not threaten the health of the coyote population. Kim Royar, the department's furbearer biologist, summed it up in a conversation we had in the spring of 2007: "Coyote hunting is perfectly acceptable, but we don't believe that the tournaments are going to accomplish what the people holding them think they're going to accomplish." John Hall, information director for the department, echoed those sentiments: "We don't believe that the organized hunts will make a difference in the long term. It doesn't do any harm to hunt coyotes. We actually encourage people to go out coyote hunting." Clearly, somebody like Jim Hoverman would differ with the assertion that coyote hunts don't do any harm, and this is a perfect example of the essential difference between the animal protectionist perspective and the wildlife management perspective: individual welfare versus population health.

The position of the state's Department of Fish and Wildlife contradicts the most important stated purpose of the coyote tournaments: killing coyotes to increase the deer population. Among wildlife managers, however, it is common knowledge that coyotes actually reproduce at a higher rate when they come under intensive attempts to control their numbers (Voigt and Berg 1987). So in the minds of wildlife biologists, all these coyote tournaments may actually increase the coyote population. Some serious coyote hunters, however, claim that the local coyote population has decreased over the last few years as hunting them with dogs has grown in popularity. When I mentioned this to a Fish and Wildlife biologist, I was told that there may well be a short-term decline in the population, but that in the long term the coyotes would bounce back in increased numbers. And as far as the deer herd is concerned, at this point the department is confident that the herd has not been adversely affected by coyotes.

Many hunters, however, including those who have never shot a coyote, just don't buy the state's position on the coyote's impact on the deer herd. Almost any hunter can reel off story after story that implicates coyotes in the serial predation of deer. They report seeing coyotes chasing and killing deer, and devouring any deer that has died by any cause, usually within about twenty-four hours. A common observation among hunters lately is if they injure a deer with a less than perfect shot and cannot locate it until the next morning, it's almost guaranteed that a coyote will have eaten it before the hunter can find it. How, they ask, can coyotes *not* be affecting the deer population when so many of their meals seem to consist of fresh venison?

To say that the Department of Fish and Wildlife does not oppose the coyote tournaments is not to say that the department's personnel are not deeply concerned about the potential of the tournaments to damage the image of hunting and thereby undermine the ethical foundation of hunting as a wildlife management tool and as a source of revenue. As Royar put it, the tournaments "may not be in the best interest of hunting." "Hunters, she said, "have to continue to be conservationists." People like Jim Hoverman see a fatal flaw in the department's response to the coyote tournament controversy. They wonder how the agency can disagree with the rationale for having the tournaments (decreasing the coyote population), but not step in and call for regulation of the tournaments. The short answer is pretty simple: Fish and Wildlife doesn't care all that much about the *motivations* for different kinds of hunting, but rather about the *effects* of hunting on game populations. So whether or not the tournaments do what the participants think they are doing is really irrelevant as long as wildlife populations don't suffer.

Still, Hoverman believes that this contradiction exposes a serious problem, even a moral one, with the way the Department of Fish and Wildlife is doing business. During both of our conversations he expressed frustration with the discrepancies in the public and private sentiments of Fish and Wildlife regarding the tournaments. He claimed that department personnel actually voiced far more concern about the hunts in private than they would ever admit in public. He went so far as to accuse the agency of "winking at de facto bounty hunts." After reviewing internal Fish and Wildlife documents his group secured under the Vermont Public Records Act, it does it appear to me that Fish and Wildlife has expressed a bit more negativity toward tournaments in private than in public.[3]

Nevertheless, by 2007 Fish and Wildlife had recommended to the Fish and Wildlife Board that a "wanton waste" provision be placed on the tournaments, requiring that a significant portion of each coyote killed in a tournament be "retained for use." The board voted the measure down, however, because they wanted to seek broader input from different constituencies, such as the fishing community. A revised wanton-waste provision has made some progress in the board review process, and several public meetings on the subject will take place. By all accounts, Fish and Wildlife is looking forward to the provision's being approved by the board. As Kim Royar stated, "As hunters we should be promoting ethical use."

Tournament Support

If we take the Department of Fish and Wildlife's word for it, the coyote tournaments will not accomplish what the participants claim they hope to accomplish. In fact, they may do something more like the opposite: increase the coyote population through intensive management. The obvious question, then, for many observers is, Why would *anybody* support these tournaments? And, as can only be expected in a society that seems to thrive on the demonization of those we perceive to be our opponents, explanations for coyote tournament support—whether from animal rights activists or other hunters—are often based on assumptions about the morality of the participants. Moral attacks, however, should not be confused with data-driven explanations for social practices. I think we can rest assured that the most beneficial aspect of the "demonization approach" is its methodological convenience. In order to explain why some people remain such passionate supporters of the tournaments without resorting to worn-out stereotypes about bloodthirsty, ignorant hunters, we must look for answers in the realm of culture—the historically anchored, materially engaged, dynamic conceptual apparatus with which groups of people make sense of their world.

Illegitimate Killers and the Necessity of Wildlife Management

A critical theme that I have consistently encountered over the years is the idea that coyotes are "illegitimate killers" of deer. I borrow this apt phrase from Garry Marvin, who claims that aristocratic English foxhunting was based on the assumption that foxes were illegitimate killers who "pose a threat to that which belongs to humans—reared game and livestock. The fox kills animals that should only be killed by humans" (2000, 205). Marvin builds his argument by explicitly drawing on Edmund Leach's 1964 article, "Anthropological Aspects of Language: Animal Categories and Verbal Abuse," in which Leach argues that human–animal relationships in England are based on categorizing animals as either *wild animals*, which act wild and live outside of human care, *domestic animals*, which behave domestically and are directly cared for by humans, or *game animals*, which act wild but are cared for by humans. Humans eat domestic animals and game animals, and wild animals should eat wild animals. But animals like foxes often eat wild, game, and domestic animals. This transgression of cultural boundaries makes them *illegitimate killers*. In other words, the fox challenges the human role in the cultural economy of killing. The same position is articulated by many Vermont hunters as they lament the damage coyotes inflict on the deer, rabbit, and sometimes even fox populations.

As we saw in chapter 6, coyotes are extremely unmanageable creatures. Having wandered into a local culture (arriving in the mid-twentieth century) that places a great emphasis on the anthropocentric management of animals and hadn't had to contemplate living beside wild canines since the nineteenth century, the transgressive coyote simply doesn't mesh well with expectations about human–animal relationships. As a result, I am hardly surprised that people aren't welcoming coyotes to Vermont with open arms, or that antiwolf sentiment also runs high among hunters in Vermont. Because of this, I question the extent to which the coyote tournaments are, in fact, contradicting age-old Vermont hunting ethics. The tournament rationale is a wildlife management one (albeit a flawed management rationale in the opinions of professional wildlife managers). As such, it is easy to see why tournament hunters aren't feeling particularly defensive about not living up to the standards of "Vermonter tradition." What's disagreed on is not the right of humans to manage wild animals; as I've noted, it's primarily the effect of coyotes on the deer population, the potential threat they pose to humans, and the seemingly wasteful killing that occurs at the tournaments. What the tournaments might truly be exposing is a point of ambivalence in traditional Vermont hunting ethics on the extent to which one animal can be wasted in order for another one to be preserved.

Cultural Politics

As I mentioned in the introduction, a strong insider/outsider discourse circulates in the state of Vermont.[4] It should come as no surprise, then, that this dichotomy saturates discussions of coyote tournaments as well (as is demonstrated by some of the newspaper letters I quoted earlier). As one adamant tournament supporter said in an interview (addressing a hypothetical antitournament protester), "What gives you the right to come into our state and destroy three hundred years of tradition?" This is an intriguing statement, since it claims that contemporary coyote tournaments, in which the prey is rarely utilized, are essentially the same as "traditional" colonial subsistence hunting, even though the first recorded killing of a coyote in Vermont occurred in 1948. To this supporter, hunting is hunting. It's that simple. It's about pursuing animals, maybe killing them, and the social bonding and reproduction of Vermont tradition that occurs within that experience. As he told me, "It's Vermont. It's what Vermont is all about."

During one of our discussions, Jim Hoverman disagreed with the claim that the tournament controversy is an uncomplicated insider/outsider issue, suggesting that it is an "insider/insider/outsider" issue. He argued that the insider/outsider argument is a rhetorical strategy employed by hunters to avoid acknowledging local opposition to the hunts and to deflect attention away from the reality of what they are doing—which, to Hoverman, is clearly immoral. While I'm not sure if tournament hunters are being quite as clever about their rhetoric and self-representation as Hoverman suggests (remember, these hunters generally didn't want any part of "utilizing" the coyotes at the tournaments I attended, even though "wanton waste" is the hot-button issue of the day), I do agree that "outsiders" are a common, culturally salient, and politically convenient opponent during discussions about tournaments, and this obscures the fact that opposition to the tournaments actually emanates from a wide variety of people—natives and flatlanders, hunters and nonhunters.

Proponents of coyote tournaments give the impression that rural Vermont is under siege from two outside invaders: coyotes and antihunting protestors (and all the cultural change the latter imply). The two handmade signs posted along the organizer's driveway during the weekend of the 2007 hunt spoke volumes on the political significance of these events. One advised anyone either not entered in the hunt or not "a 100 percent supporter of the hunt" to "get to hell off my land. Please!" and threatened that trespassers might be asked for personal identification. The other, addressed to "Antis and Protestors," announced that if they were looking for trouble, "you've come to the right place!

These tournaments are, to varying degrees for different participants, local acts of political resistance against the invasion of rural space by "antis and protestors," a literal battle with deer-hungry coyotes, and a performance of rural values and social solidarity (see Song 2000). As such, Vermont coyote tournaments are a good example of a perspective that has become axiomatic in ecological anthropology: human relationships with animals often say a great deal about, or may even be modeled after, relationships between different groups of humans (Marks 1991; Howe 1981; Mullin 1999; Descola 1992; Knight 2000a).

Throughout my research I've been told about how things are changing in Vermont. What's more significant to me than the mere existence of change (which, after all, is always occurring everywhere) is the direction of this change. While to your typical suburbanite or urbanite Addison County in November would probably seem about as pro-hunting as it gets, my informants feel that hunting is becoming less central to local culture. In January 2007 several men and women spoke with me about this issue. One Ripton woman focused on her growing discomfort with driving through town with a dead deer in the back of her pick-up. "Whereas you used to parade around with your animals and, kind of, were really proud—I mean, I think more often than not nobody leaves their tailgates down. It's like—in the truck with the tailgate up." When I asked her if she was specifically referring to the Middlebury area she said, "Oh, absolutely!" This woman's husband explained why he keeps his tailgate up: "I really do it to avoid confrontation. But also I respect the fact that some people don't like seeing dead animals. Period. And it's not something that I feel I have to show anybody."

Another couple I spoke with in January 2007 said that over "the last ten years or so" they had noticed some significant changes in local sensibilities regarding hunting. Their specific examples focused on local attitudes toward guns, explaining that in the past they would wear handguns into the local grocery store without a second thought, but not anymore.

In 1998 a teenager told me that kids who hunted were sometimes made fun of as "rednecks" at his public school in Middlebury. Perhaps the best story I've been told by a hunter about changing local perspectives on hunting actually involved this teenager. While I was visiting a Ripton deer camp in 1997, his father explained to me that one day (in the late 1980s) after shooting a deer, he decided to stop by his son's co-op preschool in East Middlebury to show him the deer. He pulled into the school driveway, walked in, and asked if his son could come outside for a minute to look at the deer in his truck. Much to this hunter's surprise, the teacher asked if the entire class could come out and look. Apparently, she thought it might make for good impromptu zoology lesson. Before this hunter knew it, his truck was overrun by little kids who wanted to investigate the deer. Though he told me the story almost ten years after it had occurred, he still seemed to be somewhat astonished by what had taken place that day. He concluded the story on a rather somber note, however, speculating that something like that would probably never happen again because most school teachers, he believed, wouldn't want their students inspecting a dead deer.

At this point, it is impossible to know if we will look back on this coyote-hunting controversy as the beginning of an era of significant change for hunting in the state. In the meantime, however, I'll be interested in seeing whether the state legislature bans the tournaments altogether, since bills were introduced in both the Senate and the House in 2005 (one by Democrat Senator Claire Ayer of Addison County and one by Republican Representative Mark Young of Orwell) to do just that. And as this book was going to press, another proposed bill to ban the tournaments was pulled from the legislative slate for 2009 because its main sponsor, Senator Ayer, felt it did not have enough Senate support.[5]

I'm also very interested in seeing how the Fish and Wildlife Board decides to handle the all-important "wanton waste" regulations that have been proposed by the Department of Fish and Wildlife. Will folks like Jim Hoverman be satisfied with such regulations or, as many hunters fear, will they just be the first small step toward the elimination of all hunting, and possibly fishing?

A third factor I'll be tracking, and one that's largely out of the control of humans, is the behavior of Vermont coyotes. What if one of them happens to attack a child walking home from the bus stop? How will that affect public sentiment regarding the tournaments? How will an incident like that affect local ideas about the "rights" of humans to kill animals? One thing I've learned through my research in Vermont is that people can make some unexpected moral decisions when wild animals get too close for comfort. It's not uncommon to hear stories about nonhunters calling their hunting neighbors to come and kill an animal that's hanging around their house. Not surprisingly, these kinds of emergency pest-control calls to local hunters become grist for the kind of "they're clueless when it comes to nature" critiques that hunters sometimes direct at nonhunters, especially flatlanders. Furthermore, many hunters are not interested in being hired guns to eliminate animals that are annoying their neighbors. One man I know who was asked by his fervently environmentalist neighbor to kill a birdseed-grubbing black bear was, frankly, a bit turned off by the request and thought the obvious solution to the problem was to take down the bird feeder.[6] After all, he reflected, what do you expect to happen when you put up a bird feeder in the Green Mountains—that only birds will use it?

In his article on English foxhunting, Garry Marvin makes an observation about rural culture that is particularly relevant to both the specific topic of Vermont coyote tournaments and a general discussion of the future of hunting in Vermont: "In comparison with urban societies, human–animal relations in rural space are complex, proximate, wide-ranging and engaged. One of the key strands of these relationships is that of the right and necessity of humans to inflict death on animals" (2000, 205). Based on my research (and, for that matter, my own youth), I believe this statement is entirely accurate.

Coyote tournaments have forced Vermonters—hunters and nonhunters alike—to think about how exactly they define this "right" to kill animals. In the process of these community deliberations, the long-standing anthropocentric approach to interacting with wild animals has been held up in comparison to more egalitarian ways of living with wildlife that stress the welfare of individual animals and question the morality of hunting. Consider, for example, the fourth- and fifth-graders who weighed in on the issue in the *Addison Independent*:

> We don't believe that the coyote hunt is fair. Coyotes are living creatures, like us. They deserve to live in peace and freedom.
>
> In addition, we are very concerned about peoples' dogs being shot and the grief it will cause.
>
> In conclusion, we think that the coyote hunts are very wrong because coyotes are living, thinking, creatures, and deserve to live in freedom.
>
> Stop the coyote hunts.
>
> (*Addison Independent*, February 20, 2006)

This is, obviously, a dramatically different approach to nonhuman animals than what has been articulated by the hunters quoted in this book. How will these young people think and feel about animals when they grow up? How will they actually experience animals? Will they still consider the hunters in their midst to be at war with wildlife, thus denying wild animals the freedom they deserve? The answers to these questions will go a long way in determining whether or not Vermont hunters, like contemporary Vermont coyotes, will someday be known as "illegitimate killers."

Hunting Paradoxes

Vermont hunters are living proof that many things in life are not what they seem to be. Viewed from afar, through the lens of centuries of American ambivalence toward hunting, these people can easily become personifications of many of the stereotypes that currently circulate in our society. As one might imagine, these stereotypes seriously distort the complexity of actual lives.

Almost anywhere I turn in my research I can point to something that an outsider to rural Vermont might describe as "paradoxical." For example, the idea of hunting in the modern world at all strikes many not only as deeply paradoxical, but also as a morally indefensible anachronism. What my research shows, however, is that hunting often serves as a culturally appropriate existential anchor heaved into the turbulent waters of modern life. What first might seem like a misplaced holdover from a bygone era eventually ends up seeming (regardless of one's personal feelings about hunting) like a fully "contemporary" activity.

Gender issues are probably even more counterintuitive than the modernity issue. Think of all the women, for example, who hunt in the face of considerable social opposition—doubt, suspicion, and even charges that they are complicit in a betrayal of feminine nature. One might reasonably assume that these women are adamant feminists who see their hunting as part of a broader political agenda, or, conversely, that they have internalized masculine sensibilities and, in effect, have been duped into feeding the flames of patriarchy. As we have seen, these kinds of assumptions just don't stand up to the testimony of actual female hunters who generally describe their activities in terms that are distinctly personal and local. What we have are stories about people who are trying to attend to their personal desires against the current of dominant cultural expectations, but, as a result, find themselves in some extremely complex relationships with other women and men.

The male hunters I've know in Vermont are no less complex than the women. As opposed to popular and academic characterizations of male hunters that tend to focus on hypermasculinity (and sometimes even misogyny), my research did not reveal any hidden hypermasculine agenda in hunting. Indeed, deer camps—the places where one might reasonably expect to find the most extreme examples of hypermasculine practices—actually proved to be more oriented around the desire to experience egalitarian and emotionally sensitive relationships between men, hardly the stereotypical characteristics of "American masculinity." And while there certainly are men in Vermont who actively oppose the inclusion of women in the hunting community, this perspective seems to be losing adherents at a fairly rapid pace.

The backbone of this study, of course, is human–animal relationships. And this may very well be the aspect of the book that calls received academic wisdom most into question. By relating to the flora and fauna around them in a manner that is both consumptive and respectful while also being conspicuously devoid of spiritual agency, Vermont hunters defy the expectations of many leading anthropologists. They force us to think more deeply about the theoretical assumptions that currently separate the Ignoble Westerner and the Noble Savage.

Extending from my discussion of human–animal relationships, I see my work as a reminder for anthropologists to pay close attention the role of material practices in the formation of cultural meanings. While this may strike the reader as a relatively modest suggestion, it seems to me that (despite the promise of "practice theory" to help us arrive at a more integrative understanding of the symbolic and material) there is a tendency among too many anthropologists to treat the "ideological" and the "material" as separate bodies of information. Most importantly for my purposes, this separation of the "things people think" from the "things people do" overlooks the importance of the processes of engagement—the ways people actually "connect," "interface," or "dialogue" with the landforms, plants, and animals in their midst—to broader cultural conceptions of human–nature relations. In the case of this particular study, the meanings of hunting for rural Vermonters cannot really be understood without understanding these practical engagements. Moreover, this division of ideas from actions helps create and perpetuate artificial boundaries between "Western" and "Indigenous" people by denying them the common ground of common practices.[1]

Most generally, this study highlights the need for more research on Euro-Americans, especially rural Euro-Americans, a group of people who presently occupy a marginalized position in the field of anthropology. If anthropologists do, in fact, as Henrietta Moore suggests, produce and subsequently rely on a fictive version of the West in their investigations, then perpetuating this Euro-American blind spot only hinders our ability to legitimately represent any of the people that anthropologists endeavor to understand. Of course, the only way we can attend to this need is to actually gather data on Euro-Americans, rather than continuing to peddle a reified, straw-man version of "Western" life.

Finally, one of my goals has been to provide a clearer image of a particular group of men and women who hunt—not in attempt to glorify them, but rather to humanize them and to grant their lives the meaning and poignancy that anthropologists have extended to so many other erstwhile "savages." As I remarked in the introduction, urban sensibilities tend to dominate environmental discourse in the United States. This is not because of some inside track from Manhattan, Washington, D.C., and San Francisco to a sacred realm of moral clarity; rather, it is merely a reflection of the power differential urban America enjoys over rural America in the so-called culture wars. I would like to think that this case study of Vermont hunters might remind us to pause before we condemn the "others" in our midst—whether they are the Middle-Eastern immigrants down the street, the lesbian couple who has volunteered to coach their child's youth soccer team, the pro-life activists on the school board, or the bearded hunters slowly approaching in a muddy pick-up on a washed-out Vermont country road.

Notes

Introduction

1. Among anthropologists, discussions of the cultural construction of nature started long before Cronon's "The Trouble with Wilderness," but not, as so often is the case, in venues or formats that would ever reach, appeal to, or influence the general public (see Ellen 1996).

2. In the case of anthropology, owing to its historic focus on non-Western cultures, studies conducted in the United States are not very common (though their numbers are increasing), particularly those involving rural white people. Michael Moffat (1992) has argued that there is actually an antirural bias in cultural anthropology.

3. I always get a kick out of contemporary American intellectuals who castigate subsistence hunters as environmental villains (*please* see White 1995).

4. I borrow the idea of a "suspect" from John Irving's novel *The World According to Garp*, in which Irving introduces the idea of "sexual suspects."

5. In chapter 2 of his *Mortal Stakes: Hunters and Hunting in Contemporary America* (2003), Dizard convincingly argues that these allegations are unwarranted.

6. While I was conducting my fieldwork a nonhunter asked me if I thought "the Southern gun culture" was the cause of the school shootings that were occurring at the time in the Southeast.

7. Only 5 percent of the people in Metropolitan Statistical Areas hunt; as opposed to 15 percent of the people in non–Metropolitan Statistical Areas, or "rural" areas (USDI 1997, 66).

8. Chittenden County, which includes the city of Burlington, is the most populous county in the state, with approximately 146,571 residents (271.9 people per square mile), according to the 2000 U.S. Census.

9. Wildlife watching, as defined in the National Survey of Fishing, Hunting, and Wildlife-Associated Recreation, includes feeding, observing, and photographing wild animals.

10. As a result of continued emigration to rural America by urbanites, low population densities have become less indicative of rural cultural sensibilities (see Fitchen 1991; Stedman and Heberlein 2001).

11. The anthropologist Janet Fitchen noticed similar perspectives in rural upstate New York, across Lake Champlain from northwestern Vermont: "Uniqueness is an article of faith, an untested assumption, in fact, an assumption that should not be questioned or tested. The ingredients of uniqueness are not always clear, yet people just 'know' that their

community is unique" (1991, 253). Stuart Marks describes a similar tendency among rural North Carolinians (1991, 84).

12. This insider/outsider perspective can be easily traced back to the state's maverick beginnings as a controversial sale of land grants that became the Republic of Vermont before joining the United States in 1791 (see Boglioli 2004).

13. The use of exclusionary, insider/outsider social references is certainly not unique to Vermont; friends of mine from rural Wisconsin often refer to vacationers from Illinois as "flatlanders." In "Measuring Modernity among Mountaineers" (2001), Susan Keefe discusses the use of the term *newcomers* in Appalachia. Janet Fitchen mentions the use of the terms *newcomers* and *outsiders* in her anthropological study of rural upstate New York (1991). Nevertheless, anecdotal evidence suggests that Vermonters take this distinction a bit further and bit more seriously than most others. In fact, comparisons to Texas for sheer state jingoism are not uncommon.

14. A critical point that must not be overlooked when considering the significance of the term *flatlander* is that urban transplants can be dire threats to local rural culture as they seek to transform Vermont into a leafier version of the places they just left. The changes that occur—such as antihunting grammar school teachers, property tax hikes, and restricting hunting on private property—often come at a cultural and economic cost to rural Vermonters. Consequently, flatlander critiques cannot be wholly understood outside of the context of urban/rural cultural and class politics. (See Jenson 2008 for the recent case of a fourth-grader being reprimanded by his teacher for talking about hunting during snack time).

15. While it flies in the face of the actual history of the state, many Vermonters view their ancestors as the white natives of a rural North American homeland. This "history," not surprisingly, depends of the denial of a permanent pre-Columbian American Indian presence in the territory we now refer to as Vermont (see Boglioli 2004 for a detailed discussion of this issue).

16. An elite private college, Middlebury draws top high-school students (who are often from wealthy families) from around the country. Many of them choose to remain in the Middlebury area after graduation. For example, Woody Jackson (the artist responsible for the cows on Ben & Jerry's ice-cream cartons) is a Middlebury graduate who grew up in central New Jersey and now lives in Addison County.

17. Figures in this paragraph and the one directly below it are taken from the following sections of the Census Bureau's *American FactFinder* Web site: Addison County, Vermont, "Population, Housing Units, Area and Density" and "Selected Social Characteristics"; Vermont, "Selected Economic Characterisitics"; and United States, "Selected Economic Characteristics." Vermont is by no means immune to rural poverty. Levels of poverty in Essex County, located in the northeastern part of the state, exceed national averages: 9.9 percent of families live below the poverty level (9.2 percent is the national average), 27.6 percent of families with a single female householder live in poverty (26.5 percent is the national average), and the median household income is $34,984. Only 10.8 percent of the population holds a bachelor's or higher degree.

18. By design, Addison County is not home to any so-called big-box stores, such as Walmart, Kmart, or Target. Though towns like Middlebury and Bristol offer a variety of specialty shops, it is rather difficult to purchase many contemporary necessities, such as basic professional attire, appliances, or electronics, in Addison County. As a result, many

people make trips to the Burlington area to shop at major national retail stores. And I can attest from my experience as an employee at a Middlebury sporting goods store that the presence of these discount retailers (in addition to mail-order purveyors like Cabela's) is taking a considerable bite out of local commerce in Addison County.

19. My doctoral dissertation research took place during an eighteen-month period from the fall of 1996 to the spring of 1998. Since that time I have made one or two trips per year back to Vermont to conduct research.

20. Although the informants were selected in large part by means of a "snowball" survey (with interviewees suggesting additional people for me to interview), I tried to diversify the population by soliciting names of hunters from a variety of people associated with different social networks. These "structured conversations" almost always took place at informants' homes, which I found quite effective since it allowed me to learn many things that would not have been apparent at a neutral location: I was able to compare their words and their actions as I witnessed their interactions with husbands or wives, friends, children, or pets. For example, I had no reason to doubt the hunter who called herself an "animal lover" after finding myself wedged on her living room couch between an enthusiastic young Doberman pinscher and a curiously hungry pot-bellied pig while trying to conduct an interview.

21. Interestingly enough, Marks holds a Ph.D. in animal ecology, not anthropology.

22. Furthermore, researchers in the anthropological subfields referred to as ethno-ecology (Gragson and Blount 1999; Nazarea 1999), symbolic ecology (for example, Nelson 1983; Ingold 1986; Milton 1993; Brightman 1993; Descola 1994; Hirsch and O'Hanlon 1995; Feld and Basso 1996; Descola and Palsson 1996) and historical ecology (Cronon 1983; Crosby 1986; Balée 1992; Balée 1998; Krech 1999) have consistently shown that culture symbolically *mediates* the relationship of humans and the nonhuman world, that human–nature relationships in a given cultural context often change over time, and that culture is not a mere epiphenomenon of a particular *optimized* mode of production (see Rappaport 1968; Harris 1977).

23. Conversely, these recent nonacademic works tend toward auto-ethnography, focusing on the experiences and philosophies of the individual authors. Richard Nelson's fine book *Heart and Blood* is an exception to this generalization, as he does seek the testimony of hunters, antihunting activists, game biologists, and others. Nelson does not deal with enough people from any specific population for his book to be considered truly ethnographic, however. In other words, it is journalistic in approach and relatively thin on ethnographic data when compared to works of contemporary academic cultural anthropology.

24. A striking exception to the reliance on quantitative methods is the work of Jan Dizard. His books *Going Wild: Hunting, Animal Rights, and the Contested Meaning of Nature* (1999) and *Mortal Stakes: Hunters and Hunting in Contemporary America* (2003) attempt to combine quantitative and qualitative methods.

1. From Extinction to Tradition: Wildlife Management

1. It is important to note that not all "local" perspectives are the same. In Vermont, for example, one may encounter people who take pride in locally grown organic food that is sold at the local co-op and practice yoga with their local instructor, or people who work in the local dairy industry and discuss their disappointment with the local deer herd at the

local gas station. Clearly, these examples represent two different, though equally strident, perspectives on the "local."

2. The Vermont Legislature meets at the Vermont State House in Montpelier for sixteen to seventeen weeks from early January to late April. The Department of Fish and Wildlife is one of many departments, divisions, boards, and councils under the jurisdiction of the Vermont Agency of Natural Resources.

3. Title 3, Chapter 51, Section 2803 provides the complete explanation of the varying powers of the legislature, the Department of Fish and Wildlife, and Fish and Wildlife Board. Perhaps most important is that both the department and the board are characterized as "advisory."

4. See Vermont Statutes, Title 10, chap. 111, sec. 4606. As far as I know, Vermont is the only state that allows the hunting of fish with firearms. Newcomers and spring visitors to the state are often confused, and mildly alarmed, when they notice people with deer rifles walking along roads or sitting in tree-stands in the marshes of Lake Champlain.

5. The legislative committee referred to by Borowske is the House Committee on Fish and Wildlife, which grants authority to the Fish and Wildlife Board.

6. Although English common law is the foundation of American game laws, modern game management was clearly not an English creation. Aldo Leopold points out that the first recorded instance of holistic game management was implemented by the thirteenth-century Mongol emperor Kublai Kahn, who not only instituted hunting seasons, but also directed elaborate winter feeding programs and habitat management (Leopold 1947, 6–7). Nearly five hundred years would pass before similar management practices were implemented in England.

7. Practically speaking, however, even in the wake of Magna Carta issues of game management were not dealt with under common law. They fell under the jurisdiction of *forest law*. It would be some four hundred years, after the execution of Charles I, before the common law/forest law distinction was dissolved (McCandless 1985, 9).

8. It was exactly this kind of rapid environmental change that led Vermont native George Perkins Marsh, in 1864, to write *Man and Nature*, which was a major catalyst—if not *the* major catalyst—of the nineteenth-century conservation movement. Today, Vermont is approximately 75 percent forested (Klyza and Trombulak 1999, 128).

9. I find it unfortunate that "native peoples" were included in this laundry list of flora and fauna, but the point certainly remains that the Western Abenaki population was greatly reduced during European colonization. Among the wildlife, caribou were also driven to extinction early in Vermont's history. And many bird, waterfowl, and fish populations were imperiled by the mid-nineteenth century. The earliest fishing regulations were instituted in 1866. Restrictions on the hunting of waterfowl and grouse were passed in 1874. Quail were protected in 1882 (Perry 1964, 6–7).

10. It is hard to say exactly how many Vermonters actually hunt deer in the state, because a generic hunting license allows one to hunt various species of "small game"—such as rabbits and grouse—in addition to deer. For example, certainly not every one of the approximately 80,000 regular license holders in 2000 hunted deer; the state's Department of Fish and Wildlife puts the figure at about 40,000. We do know, however, that 36,012 people bought archery permits to hunt only deer, and that 29,592 licenses were sold for the deer muzzle-loader season. Overall, the state estimates that about 100,000 people hunted deer in Vermont in 2000 (Vermont Fish and Wildlife Dept. 2000, 15). In that year, 20,498 deer

were killed by over 17,000 successful hunters, including over 10,000 from outside of the state.

11. By the mid-nineteenth century Vermont was fast acquiring a reputation as a good place to be from, but not a good place to live. The number of people who had emigrated from Vermont reached 200,000 by 1870 (Rebek 1982, 274). Over 50 percent of native Euro-American Vermonters were living outside the state by 1880. No other state was losing inhabitants at such a staggering rate (Morrissey 1981, 123).

12. "By 1840 there were six sheep per person in Vermont, and Addison County was raising more sheep and producing more wool in proportion to both its population and territory than any other county in the United States" (Morrissey 1981, 112).

13. There were widespread ecological benefits associated with the reforesting of the Green Mountains. As David Donath writes: "Stripped of its forest cover, upland pastures eroded rapidly and caused flooding and silting downstream. With nearly three-quarters of its landscape stripped of forest by the mid-nineteenth century, Vermont was headed for environmental trouble" (1992, 216). The transition from sheep to cows also created the open valley/forested mountain patchwork landscape that is considered essential to the state's present appeal as a popular tourist destination. As Donath states, "For every excellent Vermont view, there is a field in the foreground" (1992, 217).

14. Other methods, such as sterilization and relocation, are often popular with urban residents and have been attempted outside of Vermont with very limited success (Nelson 1997, 153, 164).

15. See Louis Warren's *The Hunter's Game* for an interesting discussion of the advent of doe seasons in early twentieth-century Pennsylvania.

16. Interestingly, the 1991 buck kill of 9,825 was the highest since 1980. Biologists claim this is the result of the many antlerless seasons in the late 1970s and 1980s. Some hunters disagree, pointing to the fact that the buck kill steadily increased from the end of the antlerless seasons in 1987 until 1991.

17. These days, the state biologists generally consider 10,000 to be a good buck kill total to shoot for, so to speak. While within the healthy winter carrying capacity of the state, it also means there are enough bucks to satisfy most hunters.

18. By *socially acceptable* I refer to that aspect of contemporary deer management that involves preventing deer from destroying hedges, gardens, and crops, and other measures that are not directly linked to the actual biological health of the local deer herd. Richard Nelson's *Heart and Blood* (1997, 159–60) provides an excellent discussion of some of the "pest control" issues involved with white-tailed deer.

19. With the increased acceptance by rural Vermonters of the scientific management recommendations of the Department of Fish and Wildlife, the debate over the antlerless seasons is not as intense as it was in the past

2. A Discourse of Interdependent Human–Nature Relations

1. To further clarify, I am not referring only to hunters. As I've mentioned, Vermont is second in the country in "wildlife watching" (USDI 2006a, 112). Of course, many hunters are probably counted among the wildlife watchers.

2. It is this general approach to the nonhuman world that makes hunting far less ethically problematic in rural areas than in more urban areas, where the nonhuman world is

often thought of as a pristine domain that should be spared the damaging effects of human exploitation.

3. I fall into line with Robert Brightman on this point. In *Grateful Prey: Rock Cree Animal–Human Relationships*, he writes: "Cree hunting is 'spiritualized' in the sense that diverse agencies—including the individual animal quarry, the game rulers, the dream guardian, and human sorcerers—are said to make rational decisions affecting the number of animals on the landscape and their accessibility to hunters. These agencies are conceived as reactive to human conduct, and any event of hunting potentially implicates a complex skein of prior events in which the hunter has participated" (1993, 186). I also want to emphasize that I am not claiming that my findings among Vermonters are generalizable across Euro-America.

4. I am not unaware of anthropological critiques, most often associated with Talal Asad (1993), that call into question the systematic view of religion championed by Clifford Geertz. While I find Asad's critique a useful one, it does not change my opinion of Vermont hunters, since I think that even Asad would require a "religious" perspective to involve a nonhuman force, entity, or essence.

5. Robert Muth and Wesley Jamison consider the growing "urban epistemology" in the United States to be a "social precursor" to the rise of the animal rights movement (2000, 846).

6. I think it also bears mentioning that the juxtaposition of rural and urban values is not something that I have had to work hard to extrapolate from my data; it is a distinction that rural Vermonters repeatedly mentioned in casual conversations and interviews. For example, in one of the last interviews I conducted during my initial fieldwork, in the spring of 1998, a man in Ripton bemoaned what he described as an ever-widening "gap" between those people who understand the material realities of life and those who do not. He deeply regretted the fact that his children were being raised at a time when so many people had completely lost touch with this knowledge.

7. A *Burlington Free Press* article, "Vt. Leads Nation in Burning Wood" (August 6, 2000), attests to the tremendous amount of wood that is utilized for heating homes in Vermont.

8. The difficulty of "living off the land" is the source of much of the respect and admiration that hunters have for the very animals they seek to kill. For example, one man called the white-tailed deer buck the "the wizard of the woods" for its ability to survive harsh weather and outwit hunters.

9. These feelings of sadness were more severe in the case of large animals (such as deer) that are perceived as more charismatic than smaller game such as grouse and rabbits.

10. Hunters did not often actually use the term *stewardship*, but they described practices that can accurately be categorized as stewardship.

11. This perspective on the balance of nature could be an example of Euro-American traditional ecological knowledge. For a thorough discussion of traditional ecological knowledge see Ellen, Parkes, and Bicker 2000.

12. This perspective has more in common with contemporary scientific ecological views than the more idealistic vision of a balanced natural world composed of harmoniously integrated plants and animals (see Scoones 1999). Also, like game biologists, the hunters I studied in Vermont were more concerned with the condition of populations—plant or animal—than the welfare of any particular member of a given species. The issue of the "natural state of nature" plays an important role in Jan Dizard's *Going Wild* (1999) as well. Our data show essentially the same trends, with hunters asserting the need to manage

nature and antihunters contending that nature will reach a state of equilibrium without human interference.

13. It's important to keep in mind that humans are the preeminent natural predator of deer: "Because humans have likely been the most important predators on deer for many thousands of years, we ourselves—and the hunger in our bellies—may have influenced the evolutionary sculpting of this creature. Selective pressures from human hunting would have favored traits such as acute senses, secretiveness, wariness and escape behaviors, camouflaging coloration, cryptic patterns of body movement, and fleetness of foot" (Nelson 1997, 15).

14. Another example of a pair of animals whose populations are somewhat dependent on each other is the moose and the white-tailed deer. Moose are thought by some to destroy habitat crucial to deer. Predictably, some hunters are very concerned about the growing moose population in Addison County.

15. Some of the hunters I talked to, however, strongly believed that the coyotes had a greater right to the deer because they are, after all, wild animals that need to eat, and unlike humans they cannot simply go to the grocery store if their hunt is unsuccessful.

16. In his analysis of English foxhunting, Garry Marvin mentions a similar trend: "It will be argued in this chapter that the fox is hunted because it kills animals—game birds and domesticated livestock—which should only be killed by human beings. The fox is a rival which competes with human interests. In terms of these legitimations the fox is perceived as a pest or vermin which needs to be controlled" (2000, 189).

17. "Coy dog" is a term used commonly in Vermont to refer to what most people call a coyote. Although the term actually refers to a coyote/domestic dog hybrid, I never understood it as such when I've heard it used in Vermont.

18. Although I chuckled to myself as I imagined a deer attacking a human being, there was no denying that this man had a serious point—a point that may greatly contribute to the appeal of hunting deer. Simply put, deer are not dangerous to human beings. A wounded deer, unlike a bear or a wild boar, will not turn on the hunter who has fired a less than fatal shot. I wonder how much this fear of coyotes is the real driving force behind the animosity toward these animals, rather than the threat they might pose to the deer population.

19. While the potential danger of coyotes to humans may not be that great, there is also no reason to dismiss the actual danger of wild animals. Anybody who has lived, or even taken a vacation, in an area with a population of large predators knows that these animals can test the limits of one's ideas about human–animal equality.

20. This new data on the hybrid nature of eastern "coyotes" disproves the idea that natural selection alone (essentially an adaptation to colder climates as the coyotes came across Canada and then south into the United States) led to larger coyotes in the eastern United States than the western states.

21. Though I have tried mightily to reconcile this perspective with the ethnographic record, I have ultimately failed. Cross-culturally it simply seems to be an untenable position. As we have learned from critiques of Ortner's (1974) proposal of a universal symbolic geometry that aligns women with nature and men with culture, we cannot blindly assume that women and natural resources are always metonymically related.

22. This sort of dim representation of Westerners and, more specifically, rural white people is certainly not unique to anthropology. It circulates widely in popular culture,

and Karl Jacoby (2001) argues that the field of environmental history has largely been a celebration of urban conservationists and a condemnation of rural culture.

23. Occidentalism refers to using stereotypes to describe "the West" in the same way that Orientalism has come to mean stereotypically representing "the East," "the Orient," or "Others" more generally. In his seminal article on the topic, James Carrier defines Occidentalism as "the essentialistic rendering of the West by Westerners" (1992, 199)

24. I am certainly not alone in critiquing the presence of the Noble Savage trope in anthropology. Most notable in more recent years has been the anthropologist/ethnohistorian Shepard Krech's controversial work *The Ecological Indian* (1999). The edited collection *Native Americans and the Environment* (Harkin and Lewis 2007) is an excellent reappraisal of Krech's book. Other particularly noteworthy articles, to name but two, are Jonathan Friedman's "Hegelian Ecology: Between Rousseau and the World Spirit" (1979) and Ernest Burch's "Rationality and Resource Use among Hunters" (1994).

25. I must admit that as an anthropologist I feel strange pointing this out, since, as a sociologist friend jokes to my wife (also an anthropologist) and me, we anthropologists are "the difference people," the ones who have relentlessly pointed out cultural difference and stressed the pitfalls of judging "other" people by "our" cultural standards.

26. See Nadasdy 2005 for an excellent example of an extreme relativist position that comes very close to eliminating any chance of meaningful cross-cultural comparison and communication. See Ingold 1996 for a detailed discussion of "engagement" in which he argues "the human condition to be that of a being immersed from the start, like other creatures, in an active, practical and perceptual engagement with constituents of the dwelt-in world" (120–21).

27. Nadasdy 2005 offers a fine analysis of the indigenous side of this story that completely fails to notice the many similarities shared with rural, white Euro-American hunters.

3. Hunting in Vermont Now

1. These choices, it is important to remember, are not "rational" choices driven by some kind of capitalist "free will," but rather are the kinds of choices that actual human beings make: ones that are embedded within particular historical and cultural contexts and are, therefore, deeply subjective.

2. This man, like a number of other hunters, thought that perhaps the appeal of hunting was embedded in our biology: "Maybe it's something left over from somewhere?"

3. A good example of people collecting wild edible plants comes from a deer camp I visited just south of the Canadian border. One of the men had collected and pickled several Mason jars of fiddleheads (immature fern fronds). After I told him how much I liked them he gave me a jar to take home. I cherished this gesture, knowing how much he liked them himself and how much time he had spent collecting and preparing them. Many people, it seems, are surprised when I tell them this story, not expecting an avid deer hunter to have any interest in searching for edible plants in the spring.

4. In doing so, she accomplished the rare feat of actually verbalizing the kinds of social pressures and desires that social theorists write about in books rarely read by nonacademics. This conversation—as well as some others—went a long way in buttressing my faith in the idea that social theory is, in fact, extremely relevant to the lives of people who are not pursuing careers in academia.

5. Jan Dizard has noticed a similar tendency in his research among hunters in Massachusetts (Dizard and Muth 2001).

6. Of course, the meanings expressed in this annual ritual are changing, and that's one of the reasons why it is so interesting to study.

7. Far from masculine posturing, the respect for the ability to carry out difficult physical tasks is wholly justifiable given the work environment in which many Vermonters exist. In his study of construction workers, Herbert Applebaum (1981) comes to the same conclusion. In situations involving manual labor, it truly is beneficial be smart, skillful, and strong.

8. Dizard (2003) also comments on the fact that hunting takes place outside the bounds of everyday life in a discussion of the reasons why hunters sometime break hunting rules.

9. Briefly, "functionalist" arguments explain social practices by the functions they provide for society or individuals, even if the participants in the practices are unaware of these supposed benefits. In the process, functionalist explanations often ignore or downplay the meaning that cultural practices have for the participants.

4. Ethics, Emotions, and Satisfactions of the Hunt

1. Of course, social theorists such as Victor Turner, Dean MacCannell, and John Fiske do illuminate the serious semiotic work of recreation, as I discussed in chapter 3. Most popular characterizations of recreation, I would argue, do not accord recreational activities the importance they deserve.

2. I think it's important to point out that an appreciation of hunting as an exciting and deeply engaging (both physically and mentally) activity is not confined to people of European descent. Even in places where it is more heavily relied on for subsistence, hunters describe hunting as an enjoyable and sometimes exhilarating activity. For example, while I was visiting an Iñupiaq (Alaskan Eskimo) camp in the Arctic Circle in the summer of 2007, a man described the "rush" he gets from whale hunting. He said it was far more exciting than bear hunting because of the sheer size and strength of whales.

3. It was also in Vermont that I realized more clearly than ever that people who hunt animals often know more about animals than the people who consider themselves advocates for that animal. I do not mean to belittle the intentions, or question the legitimacy, of advocates who do not happen to be experts on animals. I simply find this an intriguing irony. This applies to trapping as well—probably more so, in fact. Trapping is a very controversial activity, and trappers are often demonized in the popular media. Many of the hunters I met in Vermont even opposed trapping. But one thing that cannot be denied is that good trappers have an extraordinary knowledge of the habits of the animals they trap.

4. This issue of heightened awareness and extreme mental focus during hunting has been commented on by other researchers and was something I was often told about during interviews and casual conversations. Since I felt this topic is largely tangential to my particular research goals I made no attempts to investigate this topic thoroughly. There is, however, clearly a physiological/meditative aspect to hunting for many people. Furthermore, many people reported very strong physical sensations—when a deer was approaching them, for example—as a common occurrence while hunting.

5. There were a very limited number of people who hunted carnivores. In these few cases, the animals were definitely not eaten, but the hunters did not think this was ethical-

ly problematic because they did not consider these animals to be edible in the first place. The killing of these animals was always explained as a way of controlling burgeoning populations. To the great majority of people I talked to, however, killing a carnivore was not an appealing prospect precisely because it was such an extreme act of pure "sport" hunting.

5. Gender Transformations

1. I should point out that the 2 percent figure is based on a very small sample size.

2. Thomas, the first female tenured professor in the College of Natural Resources at UW–Stevens Point, has written a book called *Becoming an Outdoors-Woman: My Outdoor Adventure* (1997).

3. See the group's Web site, www.voga.org/vermont_outdoors_woman.htm

4. See www.nrahq.org/women/wot.asp.

5. Although I am confident of my overall representation of the ways male hunters treat female hunters, I do wonder if some of the rather positive statements about female hunters in my interviews are the result of my informants trying to provide responses that they thought I would like to hear.

6. Included with a Vermont hunting license is a tag that must be attached to a deer immediately after it is killed. Groups of people often hunt together and simply "fill the tags"—i.e., kill as many deer as the group has tags. This may mean that one person shoots three deer or three people shoot one deer each. This is technically illegal, but there is a wide spectrum of opinion on this issue among hunters.

7. These women receive far less scrutiny than the women who openly claim to have killed deer themselves. I would assume that this is because this open admission of bending the rules does not challenge the dominant gender ideology and, in fact, reinforces the idea that females are dependent on strong male providers.

8. A woman in Wisconsin told me that when she got her truck stuck out in the woods her primary concern was getting it out herself and not having to ask men for help for fear of the ridicule that would ensue. She eventually piled up enough branches under her tires to free her truck from the mud.

9. I am happy to report that the ownership/management of this store has dramatically changed the store's relationship with local women over the last decade and that they now offer a full line of women's clothing and regularly employ women.

10. She told me that when she first started hunting men always assumed that her husband had actually shot the deer she was bringing to the checking station. The turning point, she said, was when men began to encounter her in the woods during hunting season. As she puts it, the men were suddenly forced to deal with a startling realization: "She really does hunt!"

11. See Bronner 2005; Luke 1998; Kheel 1995; Pugh 1980; Uzendoski 1978; Slotkin 1973; see Stange 1997 for general commentary on gender and hunting; see Campbell and Bell 2000 for an interesting discussion of masculinity in rural America. For a brief, but useful, discussion of this theme see Dizard 2003, 37–39.

12. Similar to the emphasis placed on practical knowledge rather than academic knowledge often found among working class and rural populations, strength gained through actual work is often preferred over strength gained from "working out" at the gym.

6. Deer Camp

1. Several hunters I spoke to made the intriguing suggestion that the partying at deer camps is dwindling as a result of the declining participation in hunting. Their theory makes sense to me: that for people to bother going to deer camp these days they have to have at least a passing interest in actually hunting, in contrast to years gone by when "everybody" went to deer camps and they were a natural focal point for social gatherings.

2. Stuart Marks provides a brief discussion of South Carolina deer hunting clubs in *Southern Hunting in Black and White*; these deer clubs differ significantly from deer camps in size, organization, hunting techniques, and the fact that members pay annual dues (1991, 139–49).

3. This observation agrees with the findings of Nemich (1996) in his study of hunting in Oregon. The ritual celebration of one's first deer through the smearing of the slain deer's blood on the hunter's face is common in the Southeast, however, and Southeastern hunters who shoot and miss deer often have their shirttails ritually cut off in joking mockery (Wegner 2001; Marks 1991).

4. Among Vermonters, the Northeast Kingdom has a reputation as a somewhat mysterious and wild place, both geographically and socially. In a state that supports extremely high numbers of hunters, this area stands out as being perhaps the most hunter-friendly. This leads to some rather surprising habits on the part of hunters, a few of which I personally experienced. For example, one morning when I went hunting with a group of men, I was dropped off beside a dirt road and given the coordinates of where the other men would be hunting. We would be conducting a drive (described in chapter 4), and as a gesture of kindness to the visiting anthropologist, I was to be on stand, thus giving me the best chance of getting a shot at a deer. One aspect of the plan perplexed me, however: the piece of land that was being driven was across the road from where I presently stood, and shooting across or onto a road is against the law in Vermont. Feeling very ambivalent, I finally asked the guy who was dropping me off how I was supposed to deal with the road. Without a second's hesitation he said, "It's an open shot." So he drove off and there I stood waiting for a deer to pop out onto the road for my "open shot." *Only in the Kingdom*, I thought to myself.

7. Illegitimate Killers

1. Of course, most contemporary cultural anthropologists do not think that human culture is actually systematic in the literal sense of the word. Numerous critiques have clearly demonstrated the problems with assuming that cultures (whether one is referring to the "groups of people" or the "worldview of a particular group of people" sense of *culture*) operate in a truly mechanical fashion. Nevertheless, if one keeps the caveats in mind, the word *system* can be a useful metaphor for describing what's going on among groups of people who interact on a regular basis and share significant ideological common ground.

2. Many people are quite worried that these tournaments could actually lead to some sort of state regulation that would affect fishing tournaments.

3. These materials included an internal memorandum, a letter draft, and meeting minutes. The minutes clearly indicate that the possibility of a ban on competitive coyote tournaments was discussed within the department.

4. Perhaps the most well-known recent manifestation of this theme was the debate that erupted over the Civil Union bill of 2000, eventually leading to the Take Back Vermont movement—a movement that went far beyond opposing civil unions and is widely agreed upon to be a general rebellion against many "liberal" policies and perspectives that were seen as being foisted upon traditional Vermonters by flatlanders.

5. Personal communication from State Senator Claire Ayer, April 9, 2009.

6. In his article "Culling Demons: The Problem of Bears in Japan," John Knight (2000b) shows just how unpredictable and unstable public sentiment toward both animals and hunters can be.

Hunting Paradoxes

1. While I certainly do not suggest that the way one engages one's physical surroundings will, in the last instance, determine one's perspectives on animals, I think it is safe to assume that our material engagements do influence our perspectives on animals. For example, there is no denying the contrasting sensibilities regarding "nature" and animals that we observe in rural versus urban areas.

Works Cited

Abu-Lughod, Lila. 1990. "The Romance of Resistance: Tracing Transformations of Power through Bedouin Women." In *Beyond the Second Sex: New Directions in the Anthropology of Gender*, ed. Peggy R. Sanday and Ruth G. Goodenough, 313–37. Philadelphia: University of Pennsylvania Press.

Adams, William Y. 1998. *The Philosophical Roots of Anthropology*. Stanford: CSLI Publications.

Altherr, Thomas Lawson. 1976. "'The Best of All Breathing': Hunting as a Mode of Environmental Perception in American Literature and Thought from James Fenimore Cooper to Norman Mailer." PhD diss., Ohio State University.

The American Heritage College Dictionary. 1993. Boston: Houghton Mifflin.

Applebaum, Herbert. 1981. *Royal Blue: The Culture of Construction Workers*. New York: Holt, Rinehart and Winston.

Asad, Talal. 1993. *Genealogies of Religion*. Baltimore: Johns Hopkins University Press.

Balée, William. 1992. "People of the Fallow: A Historical Ecology of Foraging in Lowland South America." In *Conservation of Neotropical Forests*, ed. K. Redford and C. Paddoch, 35–57. New York: Columbia University Press.

———, ed. 1998. *Advacnces in Historical Ecology*. New York: Columbia University Press.

Bashline, Jim. 1982. "The Great American Deer Camp." *Field and Stream*, November, 55–57.

Bird-David, Nurit. 1993. "Tribal Metaphorization of Human–Nature Relatedness: A Comparative Analysis." In Milton 1993, 112–25. London: Routledge.

Boglioli, Marc. 2000. "Civil Conflict and Savage Unity: Cross-Cultural Assumptions in Ecological Anthropology." *Anthropology of Work Review* 21, no. 2: 18–21.

———. 2004. "A Matter of Life and Death: A Cultural Analysis of Hunting in Rural Vermont." PhD diss., University of Wisconsin–Madison.

Brightman, Robert. 1993. *Grateful Prey: Rock Cree Human–Animal Relationships*. Berkeley: University of California Press.

———. 1996. "The Sexual Division of Foraging Labor: Biology, Taboo, and Gender Politics." *Comparative Studies in Society and History* 38, no. 4: 687–729.

Bronner, Simon J., ed. 2005. *Manly Traditions: The Folk Roots of American Masculinities*. Bloomington: University of Indiana Press.

Burch, Ernest S. 1994. "Rationality and Resource Use among Hunters." In *Circumpolar Religion and Ecology: An Anthropology of the North*, ed. Takashi Irimoto and Takako Yamada, 163–85. Tokyo: University of Tokyo Press.

Campbell, Hugh, and Michael M. Bell. 2000. "The Question of Rural Masculinities." *Rural Sociology* 65, no 4: 532–46.

Carrier, James. 1992. "Occidentalism: The World Turned Upside Down." *American Ethnologist* 19, no. 2: 195–212.

Cartmill, Matt. 1993. *A View to a Death in the Morning: Hunting and Nature through History*. Cambridge: Harvard University Press.

———. 1995. "Hunting and Humanity in Western Thought." *Social Research* 62, no 3: 773–87.

Chamberlaine, Lee. 1981. "The Deer Camp." *The Conservationist* 36, no. 3 (November–December): 13–15.

Ching, Barbara, and Gerald Creed, eds. 1997. *Knowing Your Place: Rural Identity and Cultural Hierarchy*. New York: Routledge.

Clifford, Frank. 1999. "Hunting Camps Take Wild Out of Wilderness Experience." *New Orleans Times-Picayune*, December 1.

Collard, Andreé, and Joyce Contrucci. 1989. *Rape of the Wild: Man's Violence against Animals and the Earth*. Bloomington: Indiana University Press.

Cordell, H. Ken, Barbara McDonald, J. Alden Briggs, R. Jeff Teasley, Robert Biesterfeldt, John Bergstrom, and Shela Mou. 1995. *The National Survey on Recreation and the Environment*. The Interagency National Survey Consortium, coordinated by the USDA Forest Service, Outdoor Recreation, Wilderness, and Demographics Trends Research Group, Athens, Ga., and the Human Dimensions Research Laboratory, University of Tennessee, Knoxville.

Cronin, Edward W., Jr. 1979. "Doe Harvest: Managing the Deer Herd." *Country Journal*, November, 100–109.

Cronon, William. 1983. *Changes in the Land: Indians, Colonists, and the Ecology of New England*. New York: Hill and Wang.

———. 1995. "The Trouble with Wilderness; or, Getting Back to the Wrong Nature." In *Uncommon Ground: Toward Reinventing Nature*, ed. William Cronon, 69–91. New York: W. W. Norton.

Crosby, Alfred W. 1986. *Ecological Imperialism: The Biological Expansion of Europe, 900–1900*. Cambridge: Cambridge University Press.

de Almeida, Miguel V. 1996. *The Hegemonic Male*. Providence, R.I.: Berghahn Books.

Descola, Philippe. 1992. "Societies of Nature and the Nature of Society." In *Conceptualizing Society*, ed. Adam Kuper, 107–26. London: Routledge.

———. 1994. *In the Society of Nature: A Native Ecology in Amazonia*. Cambridge: Cambridge University Press.

———. 1996. "Constructing Natures: Symbolic Ecology and Social Practice." In Descola and Palsson 1996, 82–102.

Descola, Philippe, and Gisli Palsson. 1996. *Nature and Society: Anthropological Perspectives*. London: Routledge.

Dizard, Jan E. 1999. *Going Wild: Hunting, Animal Rights, and the Contested Meaning of Nature*. Amherst: University of Massachusetts Press.

———. 2001. "Killling with Respect: Bad Faith or Good Intentions?" Paper presented to the Eighth Interdisciplinary Conference on Human Relations with Animals and the Natural World, University of Pennsylvania. May 6.

———. 2003. *Mortal Stakes: Hunters and Hunting in Contemporary America*. Amherst: University of Massachusetts Press.

Dizard, Jan, and Robert M. Muth. 2001. "The Value of Hunting: Connections to a Receding Past and Why These Connections Matter." *Transactions of the North American Wildlife and Natural Resources Conference* 66: 61–77.

Donath, David. 1992. "Agriculture and the Good Society: The Image of Rural Vermont." In *We Vermonters: Perspectives on the Past*, ed. Michael Sherman and Jenny Versteeg, 213–18. Montpelier: Vermont Historical Society.

Eckert, Penelope. 1989. *Jocks and Burnouts: Social Categories and Identity in the High School*. New York: Teachers College Press.

Ellen, Roy F. 1996. "The Cognitive Geometry of Nature: A Contextual Approach." In Descola and Palsson 1996, 103–23.

Ellen, Roy, Peter Parkes, and Alan Bicker, eds. 2000. *Indigenous Environmental Knowledge and its Transformations: Critical Anthropological Perspectives*. Amsterdam: Harwood Academic Publishers.

Fegan, Brian. 1986. "Tenants' Non-Violent Resistance to Landowner Claims in a Central Luzon Village." *Journal of Peasant Studies* 13, no. 2: 87–106.

Feld, Steven, and Keith Basso, eds. 1996. *Senses of Place*. Sante Fe: SAR Press.

Fiske, John. 1989. *Reading the Popular*. Boston: Unwin Hyman.

Fitchen, Janet. 1991. *Enduring Spaces, Enduring Places: Change, Identity, and Survival in Rural America*. Boulder, Colo.: Westview Press.

Foley, Douglas E. 1990. *Learning Capitalist Culture*. Philadelphia: University of Pennsylvania Press.

Foote, Leonard E. 1944. *A History of Wild Game in Vermont*. 2nd ed., rev. State Bulletin, Pittman-Robertson series no. 11. Federal Aid in Wildlife Restoration project no. 1-R. Montpelier: Vermont Fish and Game Service.

Friedman, Jonathan. 1979. "Hegelian Ecology: Between Rousseau and the World Spirit." In *Social and Ecological Systems*, ed. P. C. Burnham and R. F. Ellen, 253–70. London: Academic Press.

Gilbert, Frederick F., and Donald G. Dodds. 1987. *The Philosophy and Practice of Wildlife Management*. Malabar, Fla.: Robert E. Krieger.

Godelier, Maurice. 1986. *The Making of Great Men: Male Domination and Power among the New Guinea Baruya*. Cambridge: Cambridge University Press.

Golec, Matt. 1999. "Sister in the Outdoors." *Burlington Free Press*, July 1.

Gragson, Ted L., and Ben G. Blount, eds. 1999. *Ethnoecology*. Athens: University of Georgia Press.

Gutmann, Matthew. 1996. *The Meanings of Macho: Being a Man in Mexico City*. Berkeley: University of California Press.

Gwizdz, Bob. 1999. "Women Becoming More a Part of the Hunting Scene." Newhouse News Service. November 18.

Halle, David. 1984. *America's Working Man: Work, Home, and Politics among Blue-Collar Property Owners*. Chicago: University of Chicago Press.

Harkin, Michael, and David R. Lewis, eds. 2007. *Native Americans and the Environment: Perspectives on the Ecological Indian*. Lincoln: University of Nebraska Press.

Harris, Marvin. 1977. *Cannibals and Kings: The Origins of Cultures*. New York: Random House.

Heberlein, Thomas A. 1987. "Stalking the Predator: A Profile of the American Hunter." *Environment* 29, no. 7: 6–11, 30–33.

———. 2001. Foreword. In Wegner 2001.

———. 2002. "Saving the Herd by Risking the Hunt: Human Dimensions of Chronic Wasting Disease in Wisconsin." Lecture presented at the Department of Natural Resource Recreation and Tourism, Colorado State University, Fort Collins, Colo. December 12.

Heberlein, Thomas A., Goran Ericsson, and Kai-Uwe Wollscheid. 2002. "Correlates of Hunting Participation in Europe and North America." *Zeitschrift für Jagdwissenschaft* 48, supplement 1: 320–26.

Heberlein, Thomas A., and E. J. Thomson. 1991. "Socio-Economic Influences on Declining Hunter Numbers in the United States, 1977–90." In *Transactions of the XXth Congress of the International Union of Game Biologists*, part 2, ed. Sandor Csanyi and Jozsef Ernhaft, 699–705. Gödöliö, Hungary: University of Argricultural Sciences.

Heller, Rit. 2000. "Deer Camp." *Whitetail Bowhunter*, special annual issue of *Bowhunter*, 46–48, 115.

Hennessey, Tom. 2001. "Maine's Deer Camps Were Cultural Colleges." *Bangor Daily News*, November 3.

Herdt, Gilbert. 1981. *Guardians of the Flutes: Idioms of Masculinity*. New York: McGraw Hill.

Herman, Daniel. 2001. *Hunting and the American Imagination*. Washington: Smithsonian Institution Press.

Herzfeld, Michael. 1985. *The Poetics of Manhood: Contest and Identity in a Cretan Mountain Village.* Princeton: Princeton University Press.

Hill, Gene. 1998. "Everycamp." *Field and Stream*, November, 18.

Hirsch, Eric, and Michael E. O'Hanlon. 1995. *The Anthropology of Landscape: Perspectives on Place and Space.* Oxford: Clarendon Press.

Hirschfeld, Peter. 2007. "Genetic Test Confirms Wolf Shot in Vermont." *Barre-Montpelier Times-Argus*, October 10.

Howe, James. 1981. "Fox Hunting as Ritual." *American Ethnologist* 8, no. 2: 278–300.

Ingold, Tim. 1986. *The Appropriation of Nature: Essays on Human Ecology and Social Relations.* Manchester: Manchester University Press.

———. 1996. "Hunting and Gathering as Ways of Perceiving the Environment." In *Redefining Nature: Ecology, Culture and Domestication*, ed. Roy Ellen and Katsuyoshi Fukui, 117–55. Oxford: Berg.

Irving, John. 1978. *The World According to Garp.* New York: Ballantine.

Jackson, Robert, and Robert C. Norton. 1980. "Phases: The Personal Evolution of the Sport Hunter." *Wisconsin Sportsman* 9, no. 6 (November–December): 17–20.

Jacoby, Karl. 2001. *Crimes against Nature: Squatters, Poachers, Thieves, and the Hidden History of American Conservation.* Berkeley: University of California Press.

Jager, Ronald. 1990. *Eighty Acres: Elegy for a Family Farm.* Boston: Beacon Press.

James, Allison. 1993. "Eating Green(s): Discourses of Organic Food." In Milton 1993, 203–16.

Jensen, Dennis. 2007. "Management of Deer Herd Gets Big 'Yes.'" *Rutland Herald*, March 18.

———. 2008. "Parents: Teacher Silenced Son on Hunting." *Rutland Herald*, June 20.

Jones, Robert F. 2000. "Brothers of the Wolf." *Sports Afield*, November, 66–69.

Keefe, Susan Emily. 2001. "Measuring Modernity among Mountaineers." Paper presented at the annual meeting of the American Anthropological Association, Washington, D.C. December 2.

Kellert, Stephen. 1976. "Perceptions of Animals in American Society." *Transactions of the North American Wildlife and Natural Resources Conference* 41: 533–44.

Kerasote, Ted. 1993. *Bloodties: Nature, Culture and the Hunt.* New York: Kodansha.

Kheel, Marti. 1995. "License to Kill: An Ecofeminist Critique of Hunters' Discourse." In *Animals and Women: Feminist Theoretical Explorations*, ed. Carol J. Adams and Josephine Donovan, 85–125. Durham, N.C.: Duke University Press.

Kimmel, Michael. 1996. *Manhood in America: A Cultural History.* New York: The Free Press.

Klyza, Christopher M., and Stephen C. Trombulak. 1999. *The Story of Vermont: A Natural and Cultural History.* Hanover, N.H.: University Press of New England.

Knauft, Bruce M. 1996. *Genealogies for the Present in Cultural Anthropology*. New York: Routledge.

Knight, John, ed. 2000a. *Natural Enemies: People–Wildlife Conflicts in Anthropological Perspective*. London: Routledge.

———. 2000b. "Culling Demons: The Problem of Bears in Japan." In Knight 2000a, 145–69.

Krech, Shepard. 1999. *The Ecological Indian*. New York: W. W. Norton.

Lash, Scott. 1990. *Sociology of Postmodernism*. London: Routledge.

Leopold, Aldo. 1947. *Game Management*. New York: Scribners.

Lepowsky, Maria A. 1993. *Fruit of the Motherland: Gender in an Egalitarian Society*. New York: Columbia University Press.

Lopez, Barry H. 1978. *Of Wolves and Men*. New York: Scribners.

Luke, Brian. 1998. "Violent Love: Hunting, Heterosexuality, and the Erotics of Men's Predation. *Feminist Studies* 2, no. 3: 627–53.

MacCannell, Dean. 1976. *The Tourist: A New Theory of the Leisure Class*. New York: Schocken.

Marks, Stuart A. 1991. *Southern Hunting in Black and White: Nature, History, and Rituals in a Carolina Community*. Princeton: Princeton University Press.

Marvin, Garry. 2000. "The Problems of Foxes: Legitimate and Illegitimate Killing in the English Countryside." In Knight 2000a, 189–211.

McCandless, Robert G. 1985. *Yukon Wildlife: A Social History*. Edmonton: University of Alberta Press.

McIntyre, Thomas. 2001. "Making Camp." *Sports Afield*, September, 106.

Miller, John. 1992. *Deer Camp: Last Light in the Northeast Kingdom*. Cambridge: MIT Press.

Milton, Kay, ed. 1993. *Environmentalism: The View from Anthropology*. New York: Routledge.

Moffat, Michael. 1992. "Ethnographic Writing About American Culture." *Annual Review of Anthropology* 21: 205–29.

Moore, Henrietta L. 1999. *Anthropological Theory Today*. Cambridge: Polity Press.

Morgan, David H. J. 1992. *Discovering Men*. London: Routledge.

Morrissey, Charles T. 1981. *Vermont: A Bicentennial History*. New York: W. W. Norton.

Mullin, Molly H. 1999. "Mirrors and Windows: Sociocultural Studies of Human–Animal Relationships." *Annual Review of Anthropology* 28: 201–24.

Murphy, Yolanda, and Robert Murphy. 1974. *Women of the Forest*. New York: Columbia University Press.

Muth, Robert M., and Wesley V. Jamison. 2000. "On the Destiny of Deer Camps and Duck Blinds: The Rise of the Animal Rights Movement and the Future of Wildlife Conservation." *Wildlife Society Bulletin* 28, no. 4: 841–51.

Nadasdy, Paul. 2005. "Transcending the Debate over the Ecologically Noble Indian: Indigenous Peoples and Environmentalism." *Ethnohistory* 52, no. 1: 291–331.

———. 2007. "The Gift in the Animal: The Ontology of Hunting and Human–Animal Sociality." *American Ethnologist* 34, no. 1: 25–37.

Nazarea, Virginia D. 1999. *Ethnoecology*. Tucson: University of Arizona Press.

Nelson, Richard. 1983. *Make Prayers to the Raven*. Chicago: University of Chicago Press.

———. 1997. *Heart and Blood: Living with Deer in America*. New York: Vintage.

Nemich, John G. 1996. "Deer Hunting as a Folkloric Activity in the North Coast of Oregon: Typology and Initiation-Maturation." PhD diss., Ohio State University.

Ortega y Gasset, José. 1985. *Meditations on Hunting*. New York: Charles Scribners' Sons.

Ortner, Sherry. 1974. "Is Female to Male as Nature Is to Culture?" In *Woman, Culture and Society*, ed. M. Rosaldo and L. Lamphere, 67–87. Stanford: Stanford University Press.

———. 1984. "Theory in Anthropology Since the Sixties." *Contemporary Studies in Society and History* 26, no. 1: 126–66.

———. 1995. "Resistance and the Problem of Ethnographic Refusal." *Comparative Studies in Society and History* 37, no. 1: 173–93.

Perry, Florence J. 1964. *Progress Report of the Vermont Fish and Game Department*. Montpelier: Vermont Fish and Game Department.

Petersen, David, ed. 1996. *A Hunter's Heart: Honest Essays on Blood Sport*. New York: Henry Holt.

Posewitz, Jim. 1994. *Beyond Fair Chase*. Helena, Mont.: Falcon Press.

Pugh, David G. 1980. "History as an Expedient Accommodation: The Manliness Ethos in Modern America." *Journal of American Culture* 3, no. 1: 53–65.

Pyne, Lawrence. 1996. "Deer Hunting in Vermont Turns 100." *Vermont Outdoors*, October, 16–17.

Quinnett, Paul G. 1994. "Care and Feeding of Greenhorns." *Field and Stream*, May, 28–29.

Rappaport, Roy A. 1968. *Pigs for the Ancestors*. New Haven: Yale University Press.

Rebek, Andrea. 1982. "The Selling of Vermont." In *In a State of Nature: Readings in Vermont History*, ed. Nicholas H. Muller and Samuel B. Hand, 273–82. Montpelier: Vermont Historical Society.

Regan, Tom. 1983. *The Case for Animal Rights*. Berkeley: University of California Press.

Riley, Bob. 2007. "Governor Riley Asks Citizens to Pray for Rain." Press release, Office of the Governor, Montgomery, Ala. June 28.

Robb, Bob. 1999. "Camp Idiot." *Field and Stream*, September, 71–72.

Sajna, Mike. 1990. *Buck Fever: The Deer Hunting Tradition in Pennsylvania*. Pittsburgh: University of Pittsburgh Press.

Sanday, Peggy R. 1990. *Fraternity Gang Rape: Sex, Brotherhood, and Privilege on Campus*. New York: New York University Press.

Scoones, Ian. 1999. "New Ecology and the Social Sciences: What Prospects for a Fruitful Engagement?" *Annual Review of Anthropology* 28: 479–507.

Seamans, Roger. 1947. *The Time Is Now!—A Pictorial Story of Vermont's Deer Herd*. State Bulletin, Pittman-Robertson series no. 15. Federal Aid in Wildlife Restoration project no. 1-R. Montpelier: Vermont Fish and Game Service.

Shaw, James H. 1985. *Introduction to Wildlife Management*. New York: McGraw-Hill.

Shealey, Tom. 2000. "Wildlife Myths and Manners." *Backpacker*, October, 13.

Sherwood, Morgan. 1981. *Big Game in Alaska: A History of Wildlife and People*. New Haven: Yale University Press.

Singer, Peter. 1990. *Animal Liberation*. New York: New York Review of Books.

Slotkin, Richard. 1973. *Regeneration through Violence: The Mythology of the American Frontier*. Middletown, Conn.: Wesleyan University Press.

Smith, David M. 1998. "An Athapaskan Way of Knowing: Chipewyan Ontology." *American Ethnologist* 25, no. 3 (August): 412–32.

Song, S. Hoon. 2000. "The Great Pigeon Massacre in a Deindustrializing American Region." In Knight 2000a, 212–28.

Stange, Mary Zeiss. 1997. *Woman the Hunter*. Boston: Beacon Press.

Stedman, Richard C., and Thomas A. Heberlein. 2001. "Hunting and Rural Socialization: Contingent Effects of the Rural Setting on Hunting Participation." *Rural Sociology* 66, no. 4: 599–617.

Sutton, Mark Q. 2000. *An Introduction to Native North America*. Boston: Allyn and Bacon.

Swan, James A. 1995. *In Defense of Hunting*. New York: Harper Collins.

Thomas, Christine. 1997. *Becoming an Outdoors-Woman: My Outdoor Adventure*. Helena, Mont.: Falcon Press.

Titcomb, John W. 1898. "Game in Vermont." *The Vermonter* 4, no. 4 (November): 61–64.

Tober, James. 1981. *Who Owns the Wildlife? The Political Economy of Conservation in Nineteenth-Century America*. Westport, Conn.: Greenwood Press.

Turner, Victor. 1982. *From Ritual to Theatre: The Human Seriousness of Play*. New York: Performing Arts Journal Publications.

Ucko, Peter J. 1988. Foreword. In *What Is an Animal?*, ed. Tim Ingold. London: Unwyn Hyman.

U.S. Census Bureau. 1998. *Statistical Abstract of the United States*. Washington, D.C.: Government Printing Office.

U.S. Census Bureau. 2000a. *Statistical Abstract of the United States*. Washington, D.C.: Government Printing Office.

U.S. Census Bureau. 2000b. *American FactFinder*. factfinder.census.gov.

USDI. 1993. *1991 National Survey of Fishing, Hunting, and Wildlife-Associated Recreation*. U.S. Department of the Interior, Fish And Wildlife Service, and U.S. Department of Commerce, U.S. Census Bureau.

USDI. 1997. *1996 National Survey of Fishing, Hunting, and Wildlife-Associated Recreation*. U.S. Department of the Interior, Fish And Wildlife Service, and U.S. Department of Commerce, U.S. Census Bureau.

USDI. 2002. *2001 National Survey of Fishing, Hunting and Wildlife-Associated Recreation*. U.S. Department of the Interior, Fish And Wildlife Service, and U.S. Department of Commerce, U.S. Census Bureau.

USDI. 2006a. *2006 National Survey of Fishing, Hunting, and Wildlife-Associated Recreation*. U.S. Department of the Interior, Fish And Wildlife Service, and U.S. Department of Commerce, U.S. Census Bureau.

USDI. 2006b. *2006 National Survey of Fishing, Hunting, and Wildlife-Associated Recreation—Vermont*. U.S. Department of the Interior, Fish And Wildlife Service and U.S. Department of Commerce, U.S. Census Bureau.

Uzendoski, Emily Jane. 1978. "Deer Hunting in Popular Outdoor Magazines." *Journal of American Culture* 1: 606–15.

Vanden Brook, Tom. 1991. "Men, Keep Out: Workshop Opens Nature's Door to Women." *Milwaukee Journal*, September 15.

van Zwoll, Wayne. 1998. "Old Plaid Flannel." *Field and Stream*, August, 100–102.

VDTM. 2007. *The Travel and Tourism Industry in Vermont: A Benchmark Study of the Economic Impact of Visitor Expenditures on the Vermont Economy—2007*. Vermont Department of Tourism and Marketing.

Vermont Statutes. Vermont Statutes Online: www.leg.state.vt.us/statutes.

VFWD. 2000. *A Report of Vermont's 2000 Deer Hunting Season*. Montpelier: Vermont Fish and Wildlife Department.

VFWD. 2008. *2008 Vermont White-Tailed Deer Report*. Montpelier: Vermont Fish and Wildlife Department.

Voigt, Dennis R., and William E. Berg. 1987. "Coyote." In *Wild Furbearer Management and Conservation in North America*, ed. Milan Novak, 345–57. Ontario: Ontario Trappers Association under the authority of the Ontario Ministry of Natural Resources.

Warren, Karen J., ed. 1997 *Ecofeminism: Women, Culture, Nature*. Bloomington: University of Indiana Press.

Warren, Louis. 1997. *The Hunter's Game: Poachers and Conservationists in Twentieth-Century America*. New Haven: Yale University Press.

Wegner, Robert. 2001. *Legendary Deer Camps*. Iola, Wisc.: Krause Publications.

White, Richard. 1995. "'Are You an Environmentalist or Do You Work for a

Living?': Work and Nature." In *Uncommon Ground: Toward Reinventing Nature*, ed. William Cronon, 171–85. New York: W. W. Norton.

Wikan, Unni. 1990. *Managing Turbulent Hearts*. Chicago: University of Chicago Press.

Williams, Joy. 1990. "The Killing Game." *Esquire*, October, 113–28.

Willis, Paul. 1981. *Learning to Labor*. New York: Columbia University Press.

Wilson, P. J., W. J. Jakubas, and S. Mullen. 2004. *Genetic Status and Morphological Characteristics of Maine Coyotes as Related to Neighboring Coyote and Wolf Populations*. Final Report to the Maine Outdoor Heritage Fund Board, Grant 011-3-7. Bangor: Maine Department of Inland Fisheries and Wildlife.

Index

MARC BOGLIOLI received his Ph.D. from the University of Wisconsin–Madison and is currently an assistant professor of anthropology at Drew University in Madison, New Jersey. He lives in Pottersville, New Jersey, with his wife, Joslyn Cassady, and his daughters, Willa and Quinn.